Cold Blue

Verne Nobles, Sr.
William A. Blum

Acknowledgment

In writing this book, I've always held great esteem for the pilots who fought in World War ll. It was a time when young boys became men in our world war struggle to defeat tyranny. When my partner and I decided to explore many of the strange and unexplained stories that came out of World War II, we selected a Navy Airman to be the catalyst who would bring one of the strangest tails of secret experiments that has, to this day, lingered unanswered in the pages of that time period. With this in mind, we look to the Airmen who flew in both the European and South Pacific theaters and honor over 65,000 pilots who fought a world war that gave us the very freedoms we so cherish in our lives today.

William A. Blum

Author's Dedication

During my sixty-five years working in the entertainment industry, there have been those who came along, reached out, and gave a kid from the east side streets of Detroit a vision of a magical world far from the inner-city block that was his total domain as a young boy. Many of their names will mean little to the majority of those who most kindly have chosen this book to read; however, each of them offered a path, a stepping stone that gave me the courage to face my impossible journey from the East side streets of Detroit to the studios of old Hollywood as a writer, producer, director, and most recently in my aging years, an author whose passion most recently matched his creativity of a lifelong landscape artist.

Verne Nobles, Sr.

Author's Dedication

The most meaningful accomplishment and joy of my life has been to witness the birth of my children and watch them grow up. One always wishes and hopes that their children will have a better life than they were served, but the twists and turns of destiny sometimes bring sadness. Philip, my son, passed away far too soon, but during all that darkness, my daughter was the bright light in my life. To this day, she encourages me to turn to happiness in all my ambitions. With her encouragement, I endeavored to attempt yet another goal as an author and have become somewhat successful. My dearest daughter Trudy, this book is dedicated to your everlasting love.

William A. Blum

About the Authors

In 1979, Verne Nobles Sr. was invited to leave his commercial production company in Detroit, Michigan, after producing and directing over 3200 national and international commercials and join 20th Century Fox. Within two years, Verne had written, produced, and directed three prestigious, wide-screen documentaries for the studio, featuring narration by Orson Welles, Lowell Thomas, and Charlton Heston, with guest appearances by Alex Haley, President Gerald Ford, Louis L'Amour, Dick Van Dyke, Gregory Peck and Margaret Truman, as well as other historical figures. These films were followed by co-producing the television production of Louis L'Amour's "The Shadow Riders" and L'Amour's "Shaughnessy", which would be the beginning of a long 65-year journey in the various aspects of the entertainment industry.

About the Authors

William A Blum… There were times when I would sit up in my tower crane two hundred feet in the air, like an eagle and gaze over the landscape. I fixed my eyes on the people below me and wondered what life they had intended for themselves. Construction is one of the most challenging careers with respect to a physical attribute. When I entered this field, I was very young, filled with aspirations and desires to become something more than just a person who operated heavy equipment. I've had several successful businesses and enjoyed the fruits of my labor. During that time, I enjoyed a variety of encounters, but the most difficult of all the tasks I placed before me was sitting down with a pen and paper, creating something people would actually read and find enjoyable.

Author's Photograph by Richard Kelly

Table of Contents

Introduction

Within the World War two history books, there remain intriguing accounts of secret government missions, which remain a mystery. One such story begins on a small, uninhabited island in the Philippines in 1939. A group of American military construction workers and Navy intelligence officers arrive there with top-secret instructions to build a small aircraft runway on the northwest section of the island. Little do they know, they're about to discover something extraordinary hidden deep beneath the island's surface. Geophysics believed that the original island arcs were formed millions of years ago by tectonic volcanic eruptions at the southern part of the Philippine tectonic sea plate. Volcanic, lava tubs, and cavern formations were not unusual in this part of the Philippines. However, the initial seismic soundings and other tests revealed a strange electron surge. Further tests discovered that these surges were coming from a large, deep lava cavern. The next series of tests uncovered a unique phenomenon that simple physics could not explain. As the construction on the airfield continued, Vice Admiral Philip Collins, with headquarters, Naval Intelligence, read the island reports and found it necessary to contact Astrobiologists Professor Hiram Bigalow and Professor Augustus Franco. At the time, both were involved in a highly secret study of electromagnetics in crystal caves located in Canada. Hiram and Franco both arrived on the island site and within forty-eight hours, they continued the exploration. An excavation was planned to find a way to the caverns below. Reports continued to be sent and read by a select few intelligence officers.

In 1941, as the war in the south pacific suddenly widened, there were no further communications recorded. The last report, coded "Cold Blue," stated that in one cave alone, they stumbled upon crystal spikes giving off continuous electron surges. Naval intelligence received that report but with no further information. It was rumored that the next time the two scientists' names appeared was in 1943, at a shipyard in Philadelphia. Witnesses alleged that a destroyer escort successfully became invisible, but the ship inexplicably teleported to Norfolk, Virginia, for several minutes and then reappeared in the Philadelphia yard. There were other reports about similar experiments at about the same time. The Navy Department and a long list of alphabet intelligence agencies had firmly denied that the incident at the Philadelphia shipyard actually occurred.

Furthermore, there was no record of Bigalow and

Franco ever working for the government from 1939 through 1943. Their story ends sadly in a suspicious plane crash when it was shot down over the China Sea in March 1945. The downing of the private plane, which took the lives of the two exceptional scientists, caused the loss of the secrets to the crystals and their extraordinary power. The true story of the discovery of the crystals vanished in the downing of the aircraft leaving only one question unanswered. Did they actually find a powerful, continuous energy source produced from large, needle-like crystals found in a lava cavern deep inside a South Sea Island, or was this just another World War II historical mystery lost in time? Did Bigalow and Franco plan one last experiment? In this book, we will dive into the secrets of the code word "Cold Blue" and the amazing events that transpired in search of its unlimited power.

Chapter 1:
The Awakening

The sun peeks over the horizon, casting its golden light upon a Naval Island airfield entrance where a weathered "NO TRESPASSING" sign stands sentinel. Beneath its watchful gaze rests the U.S. Aslito Navy air base in the Marianas. The year 1944 June 17. On the tarmac, a silhouette of a Grumman F6F Hellcat, a single-crew carrier-based fighter aircraft, awaits its next mission.

"ASLITO NAVY AIR BASE," reads a sign nearby as the dark morning sky gradually illuminates the sleek, dark body of the fighter plane. Lt. Nathan Miller, a ruggedly handsome Navy pilot clad in his flying suit, harness, and parachute, strides purposefully towards the aircraft, unaware of the profound impact this flight will have on his life and the course of history.

Nearby, four men load a canvas tent into a jeep as mechanic Charlie Polansky watches, visibly perturbed. "Assholes!" he mutters, shaking his head, while Lt. Miller observes the scene, noticing Charlie's frustration.

"What was that all about?" Nathan inquires as he

approaches Charlie, who greets him with a nod. "Good morning, Lieutenant," Charlie acknowledges. "Charlie," Nathan replies, his curiosity piqued.

Charlie gestures towards a white soup bowl-looking attachment under the fuselage of the plane. "Take a look under here. Naval intelligence officers... At least that's what they told me," he explains.

Bending down to inspect the mysterious contraption, Nathan's expression turns incredulous. "What the hell is that?!" he exclaims, taken aback by the unfamiliar apparatus.

"They set up a tent and told me to take a hike. Handed me this," Charlie says, handing Nathan a sealed document. Without hesitation, Nathan opens it and begins to read.

"The only thing they would tell me is it's some kind of 360-degree camera," Charlie adds, shedding light on the situation.

Nathan furrows his brow, puzzled. "There was no mention of a camera in the briefing yesterday," he remarks, glancing at the bowl.

"Admiral Tuners' orders. It's just a test. Report to Commander Brereton Lewis. Follow the coordinates and avoid any contact with the Japs," Nathan reads aloud, absorbing the contents of the document.

As the sun continues its ascent, casting its warm glow over the airfield, Charlie removes the chalks from the plane's wheels and returns to Nathan's side.

"Well, she's ready. I found a way to stuff your sea bag behind your seat headrest," Charlie informs Nathan, his concern evident in his voice.

Nathan nods, tucking the document into his jacket pocket. "This explains why they held me back from leaving with the squadron and returning to the carrier. The word is there is a major battle coming in the next few days," he muses.

"You have to get out of here. There is also a storm moving into the southeast and headed this way," Charlie urges, his tone urgent.

"I saw the report. It shouldn't be a problem. I'll take it to 30,000," Nathan reassures him, determined to carry out his orders despite the impending weather.

Charlie Pollack's voice carried over the tarmac. "The 19th and 73rd Fighter Squadron is due in here this morning, and they're going to get the worst of the storm."

Nathan slips into the cockpit, grabs his harness,

places it around his shoulder, and snaps it shut. He flips the Magneto switch and hits the starter, the engine of the Hellcat igniting with a powerful roar. Charlie salutes, and Nathan returns the gesture.

"Good hunting, pal!" Charlie yells as the Hellcat taxis off, watched by the four men loading the jeep. One of them, camera in hand, films the action.

Alone in the open sky, oxygen mask now in place, the Hellcat climbs to 30,000 feet and levels off. Nathan glances at the instruments before scanning his surroundings, his gaze lingering on something in the distance.

Above the billowing clouds, two Japanese Mitsubishi A6M4 Zeros emerge, ascending from the dark clouds below.

In the cramped confines of the cockpit, Nathan's breath quickens as he assesses the situation. "Shit! They've spotted me!" he mutters under his breath. With swift reflexes, he rolls the Hellcat to the left, initiating a dive toward the oncoming Zeros.

As the Hellcat hurtles downward, Nathan's focus remains razor-sharp. Outside, the two Zeros climb head-on toward the descending aircraft, the tension palpable in the air.

Inside the cockpit, Nathan's finger hovers over the firing trigger on the stick, his decision made in an instant. With a decisive press, the six M2 Browning .50 cal machine guns roar to life, unleashing a barrage of firepower.

The climbing Zeros retaliate, opening fire on the Hellcat as it executes a hard maneuver to the right, narrowly evading the incoming bullets.

Nathan knows he has the advantage; the Zeros' lack of hydraulic boosting makes it difficult for them to maneuver at high speeds, giving the Hellcat the upper hand. Again, he pushes the firing button, igniting the AN/M2 cannons and sending another blast of 225 rounds at the green Zero, hitting his target.

In the heat of the moment, the other Zero streaks perilously close to the Hellcat's left wing. Nathan's gaze darts to the right, searching for the other assailant he had fired on, only to witness the Zero erupting in flames, sending shockwaves through the air.

The sudden explosion throws the Hellcat off balance, jolting Nathan's focus back to the remaining Zero, now stalled and turning, its trajectory aimed directly at him. Again, his thumb presses and his gun ignites, sending a blast at the second Zero.

Outside, amidst the chaos of combat, the Zero unleashes its deadly payload, firing its four 20mm guns at the Hellcat.

In a desperate bid for survival, Nathan maneuvers the Hellcat hard to the left. But the Zero's bullets find their mark, tearing through the engine and colliding with the canopy above.

As the Zero passes beneath the Hellcat, it detonates in a fiery explosion, hurling shrapnel that rends the lower fuselage of Nathan's aircraft. Both planes plummet towards the storm-ridden abyss below, locked in a deadly dance of fate.

Inside the cockpit of the Hellcat, Nathan's heart pounds as smoke from the engine fills the air inside the canopy, mingling with the acrid scent of cordite. He reaches for the canopy, only to realize he's locked inside the plummeting aircraft.

Desperate to maintain control, Nathan wrestles with the controls, but it's a futile struggle against the forces of nature as the falling Hellcat enters the storm. The storm rages outside, its winds howling and lightning illuminating the darkened sky.

Through Nathan's eyes, an island looms in the

distance, a distant beacon of hope amidst the chaos.

Nathan grips the stick tightly, his knuckles white as he fights to steady the descent. With a swift motion, he lowers the flaps and wheels, hoping to execute a dead stick landing on the beach below and avoid the churning ocean. As the Hellcat continues towards the island, lightning forms a sphere that engulfs the fuselage for several seconds as a cool breeze fills the cockpit, followed by an eerie blue glow that next fills the cockpit, leaving the lighting to form the electrical sphere to morph into a blue tunnel to the island below, that now rushes towards him. He shakes his head, trying to keep from passing out, and then...

The jungle below embraces the Hellcat fuselage as it crashes, its impact sending foliage scattering in all directions. The twisted wreckage lies partly buried amidst the lush vegetation, its shattered cockpit revealing Nathan's motionless form held in place by his harness.

As lightning flashes in the sky, Nathan regains consciousness inside the cockpit. Blood trickles down his face from a cut on his forehead as he surveys the darkened interior. With steady hands, he removes his goggles, cap, oxygen mask, and headset, his movements deliberate despite the chaos around him.

With a sense of urgency, Nathan fumbles for the canopy release, struggling against the broken mechanism until the cracked cover finally gives way to the misty air outside. He releases his parachute harness and attempts to free himself from the wreckage, but his efforts are in vain as he falls back into the seat.

After a moment's pause, Nathan summons his strength and pulls himself up through the narrow, broken opening, standing precariously on the seat as he peers into the wet, misty darkness beyond. Despite his best efforts to maintain his balance, his head spinning, he has difficulty holding his gaze as he loses his footing and tumbles onto the wing and then slips off.

As he lies on the wet ground, Nathan struggles to rise to his feet, each step a testament to his resilience amidst the pain. With trembling hands, he clings to nearby trees, slowly making his way through the rain and blurred darkness toward the faint lights in the distance. But as his knees buckle once more, he falls to the ground, his determination unwavering despite the odds stacked against him. He finally reaches for something and pulls himself up. The night hides the rhythmic sounds of the jungle. Lightning flashes, exposing a narrow, rain-soaked path, and he slowly steps forward. Somehow, he manages to slog through the rain-covered jungle foliage. The continuous lightning strikes illuminate the narrow path. Time has no meaning as he fights to remain conscious, pushing through the thick, wet, hanging foliage. All of a sudden, a faint light becomes his beacon, but his legs give way. He falls to his knees and tries to crawl toward the faint light.

A solitary light casts elongated shadows in the dimly lit area outside the rear door of the large three-story building. The rain falls steadily, punctuated by distant rumbles of thunder. Amidst this backdrop, a young woman emerges from the door, clutching a small flashlight tightly in her hand. With purposeful strides, she navigates down a narrow path toward a cluster of island cottages nestled at the edge of the jungle.

Approaching one of the cottages, the young woman quickens her pace, the urgency evident in her movements. The area is briefly illuminated by a flash of lightning, revealing the contours of the cottages against the night sky. With another rumble of thunder, the young woman reaches her destination, the safety of the cottage's small porch, and pauses, shining the light into the open doorway before stepping inside.

She walks inside as the narrow beam of the flashlight

illuminates the interior, revealing a figure sprawled on the floor of the small living room. A moment of silence hangs heavy in the air as the young woman stands frozen in place, her gaze fixed on the prone form before her. The light remains trained on the motionless body, highlighting the pool of blood beside his head, mixed with the rain-soaked carpet.

With a steadying breath, the young woman closes the door behind her, shutting out the relentless rain. Removing her shoulder wrap, she reaches for a nearby light. Turning it on, she kneels beside the fallen figure, her white nurse's uniform contrasting starkly against the dim light of the room. She presses two fingers to the man's neck, her expression grave as she assesses his condition. She rises and enters another room, returning in moments with a first aid box, towels, and a bowl of warm water. Reaching for the buttons at the top of his flying suit, she loosens it so it is not tight against his neck.

“Well, now, let’s see what you’ve done to yourself.”

Chapter 2:
It's Impossible

The morning light filters through the window, illuminating the small cottage room. The rainstorm has ended, leaving a clean, sweet scent of the South Seas lingering in the air. Nathan lies motionless, propped up against a cushioned couch. His flying suit and shirt are now open, his dog tags exposed. The young woman finishes placing bandages on his forehead. Beside her are cotton wraps, tape, scissors, and a small brown bottle alongside a bowl containing a blood-stained cloth. As she adjusts the bandage, Nathan suddenly wakes and grabs her wrist.

"Who are you?" he demands, his voice harsh.

"My name is Anahera, and you're hurting me," she responds calmly.

"Where am I?" Nathan's grip tightens.

"Half Moon Bay Resort."

"How many Japanese are here?!" he asks, still holding her arm tightly.

"Japanese? I think there is a young couple from

Nikko, but most here are from Australia and America," Anahera explains, relaxing her arm.

"Is this a prison camp?" Nathan's eyes narrow with suspicion.

"No, lieutenant. It is a private hotel resort. Now, may I have my arm back?"

"Some sort of military rest area?" Nathan's tone is accusatory.

"Release my arm," Anahera insists.

Nathan lets go of her arm, watching her intently. She turns and picks up her cell phone. As she turns back to him, he swiftly takes the Remington M1911A1 automatic from his shoulder holster and points it at her.

"Put it down... now!" he commands.

Anahera sets the cell phone down, and Nathan grabs it.

"What the hell is this?" Nathan demands, examining the phone.

"It's a cell phone. I was about to call the doctor. That hole in your head is more serious than I thought," Anahera explains.

“Cell phone?” Nathan repeats, perplexed. He looks at the phone, then at the computer and printer on the table.

“What are you, some kind of communication officer?” he asks, suspicion coloring his voice.

“No. I am the nurse who bandaged your head injury. Now, lieutenant, please put that gun down. I am not going to hurt you with my cell phone,” Anahera says firmly.

Nathan lowers the gun but keeps it in hand. He touches the bandages, wincing slightly, and then looks down at the bowl and brown bottle.

“You’re a nurse?” he asks, looking towards the computer.

“Yes,” Anahera confirms.

“What is that?” Nathan points at the computer, still wary.

“It’s a computer,” Anahera explains gently, seeing his confusion. “You must have hit your head pretty hard. It’s used for communication, work, and many other things.”

Nathan stares at the computer, trying to make sense of everything.

“Where exactly am I?” he finally asks, the tension in his voice softening a bit.

"You're in a safe place, lieutenant. Half Moon Bay Resort. You're not in any danger here," Anahera assures him.

Nathan looks around the room, taking in the unfamiliar surroundings, the modern technology, and Anahera's calm demeanor. Slowly, he starts to lower his guard, though the confusion and wariness remain in his eyes.

He looks again at the computer.

Anahera recognizes his mixed emotions. "It's a computer and printer."

"It's a computer and printer."

Nathan, still bewildered by all the modern technology, exclaims, "What does it do?"

Anahera hesitates, then asks, "How did you get here last night in that thunderstorm?"

"What does that thing do?" Nathan repeats more insistently.

"It's like a typewriter, but smarter. I can show you..." Anahera begins, moving cautiously.

Still holding the gun in his hand, he motions with it. "Go ahead."

Anahera, still on her knees, moves to the computer

and opens the browser.

“According to your dog tags, your name is LT. Nathan Miller, US NAVY,” she says, typing his name into the search bar.

Nathan looks down at his open flight suit to see his tags.

Anahera continues, “My god...” She turns to look at Nathan as he tries to move to see the screen but is held back by a sharp pain in his head.

“What does it say?” he demands.

“What's the date today?” Anahera asks, looking at the photo on the screen and then back at Nathan.

“The date?” Nathan is confused.

“Yes, the date.”

“June 17th or 18th. What does that thing say?”

Anahera looks at him seriously. “What year is it, Nathan?”

“The year? 1944. What the hell does that damn thing say?”

She reaches for a newspaper sitting next to the computer and hands it to Nathan. A vision of disbelief and continued confusion shrouds Nathan as he keeps gazing at the newspaper.

“Read the date on yesterday’s paper: June 17th, 2025,” Anahera instructs.

Nathan is stunned, looking between the screen and the newspaper. The gun falls from his hand, and he places both hands on his head.

Nathan looks at the paper, then at Anahera, who turns the monitor so he can read the words. She reads aloud the information on the monitor: “Lieutenant Nathan Miller, USNR, Fighting Squadron assigned to the aircraft carrier USS Lexington, is reported missing in action on June 17, 1944. His Grumman F6F Hellcat, a single-crew, carrier-based fighter aircraft, is shot down north of Leyte over Carigara Bay two days before the start of the Battle of the Philippine Sea. The air-to-sea action eliminates the Imperial Japanese Navy's ability to conduct large-scale carrier actions. It takes place during the United States' amphibious invasion of the Mariana Islands during the Pacific War.”

“Impossible. That is not right. That’s crazy!” he exclaims.

Anahera places a hand on his shoulder. “As insane as it sounds, you are Lieutenant Nathan Miller, born August

13th, 1922."

Nathan shakes his head in disbelief. "That is just nuts."

Anahera stands and helps Nathan to his feet. "Come on, let's get you cleaned up and dry," she says gently.

"Come on. Do you feel strong enough to stand?"

Nathan blinks, still trying to process everything. "What?"

"You need to get out of those wet clothes and into a hot shower. I am going up to the gift store to get you something appropriate to wear," Anahera explains.

Nathan takes a deep breath. "Give me a minute."

"There is a stool in the shower. I suggest you sit on it and put my shower cap over that wound," Anahera advises.

"No. Wait... the war?" Nathan's mind races.

"Your war ended in 1945 when the United States detonated an atomic bomb over the Japanese city of Hiroshima," Anahera says gently.

"An ATOMIC bomb?" Nathan is incredulous.

"They dropped a second atomic bomb on the Japanese city of Nagasaki," Anahera adds.

"Just two bombs?" Nathan asks, his voice filled with disbelief. He sits on the sofa stool and looks at the screen, still in a state of wonder.

"Here, I'll show you," Anahera says as she moves to the computer, her fingers working the keyboard. A picture appears, revealing the dropping of the atomic bomb on Hiroshima. "An atomic bomb, how horrible. What about Hitler?" Nathan asks, his eyes wide.

Anahera types again, and historical footage appears, showing the streets of New York filled with celebration and newspaper headlines proclaiming the end of the war.

"Hitler committed suicide in his bunker, and Germany surrendered on May 7, 1945," Anahera explains.

"That thing is amazing. I am going to give you some names to put in that thing," Nathan says. He slips off his jacket and begins to pull back the top part of his flight suit.

Anahera drops to her knees, ready to type. "Give me the names."

Nathan thinks for a moment. "Admiral Philip Turner, Commander Brereton Lewis, John..."

As Nathan says the names, Anahera begins to type. Each keystroke brings them closer to the answer. He studies the screen, then looks to Anahera. There is hesitation in his voice as he speaks the words. “Helen and Paul Miller, Detroit, Michigan.” The computer responds, displaying the images of the couple on the screen.

After a moment, Anahera asks, “Your parents?”

Nathan nods.

Chapter 3:
The Disappearing Professors

Admiral Jack Davis, a tall, lean man in his fifties, strides into the NSA Executive's Dining Room. He makes his way to a corner table where Colin Brown, the NSA Senior Intelligence Officer, stands to greet him.

"Good morning, Jack. Thanks for making the trip," Colin says, extending his hand.

"No problem. Nice to see how the other half live," Admiral Davis replies with a smile as they shake hands and sit down.

A waiter appears, carrying a coffee urn. "What may I serve you, Admiral?" he asks.

"Just coffee," Admiral Davis responds.

The waiter pours the Admiral's coffee and freshens Colin's cup.

Admiral Davis takes a sip and then looks at Colin with a warm smile.

"How's my godson doing?"

"We just found out yesterday that he has been accepted at Annapolis, thanks to you," Colin replies, his gratitude evident.

"He was prepared. I just gave him a push," Admiral Davis says modestly.

Colin takes a file from a black envelope and hands it to the Admiral. "I hope this is not a waste of your time, Jack. It popped up last night on the watch list—thirty-four out of thirty-eight markers. That South Sea operation is tagged a ghost from the war years but remains on the list for some strange reason."

The Admiral reads several pages, sips his coffee, and then closes the file. "Who else has seen this?"

"Only the agent who was on duty last night who transfers all incoming quotable anomalies. I locked it and specified for my eyes only," Colin assures him.

"The woman?" the Admiral asks.

"A nurse working at a private island resort in the South Seas. Nothing exceptional about her. However, thirty-four out of thirty-eight markers. Cannot be ignored," Colin explains.

"Maybe nothing, but let me look into it," Admiral

Davis replies thoughtfully.

"She hit all the markers, Jack. She only left out four names," Colin emphasizes. "That is what makes it interesting," the Admiral muses.

"When I saw this, I remembered you investigated these kinds of weird stories during your first year with intelligence," Colin recalls.

"Is that what my file says?" Admiral Davis asks with a raised eyebrow.

"No, you told me one night we were drinking Shanghai Pincers in Bangkok when you made Lieutenant Commander," Colin answers with a chuckle.

A beat passes as they both remember the night. "Ohhh. I could hardly talk for two days. That stuff was lethal," Admiral Davis says, shaking his head.

"Were the Leyte stories true?" Colin asks, leaning in with curiosity, "Especially this one?"

Admiral Davis takes another sip of his coffee, his expression contemplative. "Maybe," he says, leaving the question hanging in the air.

Admiral Jack Davis and Colin Brown continue their conversation in the quiet dining room. The waiter leaves them alone.

Admiral Davis begins, "The pilot, Lieutenant Nathan Miller, USNR, is pulled from his squadron, and his Grumman F6F Hellcat is equipped with a secret device supposedly used in the Philadelphia Experiment." His voice is low and serious. "The Hellcat goes missing two days before the Battle of the Philippine Sea."

"Admiral Philip Turner and Commander Brereton Lewis, CO of the top-secret projects tied to this device, are killed when a DC-4 is shot down in friendly skies four days later. The professors who created this thing are transferred to the New Mexico Manhattan Project. All of this happens in June of 1944. From that day forward, they do not show up in any of the major reports."

"I got all that from a deep dive into the original request to monitor these names seventy years ago," Colin says. "The same professors and Major who were tied to the Navy's Philadelphia Experiment and other similar events seven months earlier. Then, the missing Hellcat report. The question remains: is any of it true?"

"Maybe," Admiral Davis replies. "The problem is that the professors' names do not show up on the New Mexico roster, and there is no further record of either of them or what happens to the device... if it actually exists."

"Did you ever follow up on the rumors that they operated a science lab outside the country?" Colin asks.

"No," the Admiral admits. "As this report says, it was thought to be somewhere in the Philippines. I never found any sign of it. Wrote it off as another strange war story."

"I will keep an eye on our nurse," Colin promises. "Oh, and I did a satellite pass over the nurse's island. No military aircraft found."

"Good," Admiral Davis says, finishing his coffee. "Let's proceed cautiously. We don't want to tip our hand too soon."

Colin nods, understanding the gravity of the situation. "I'll keep you updated."

As they stand to leave, Admiral Davis puts a reassuring hand on Colin's shoulder. "We've seen stranger things over the years. Keep digging. We might be onto something significant."

Chapter 4:
A Whole New World with The Same Old Secrets

Anahera approached the cottage, struggling under the weight of several bags. With a push, she opened the door, stepped inside, and put the packages down.

Nathan, now dressed in casual island attire, sets at the computer. He has a yellow pad in his hands and was jotting down some notes. "Are you still at the computer?" Anahera inquired. Nathen looked up. "I think I am finally getting the hang of it," he said as his attention returned to the screen.

"My God," Nathan muttered, his eyes fixed on the screen, "the world did not learn a damn thing after 1945 and the bomb. Korea, Vietnam, Afghanistan, Russia, China... the political wars just keep repeating the same old bullshit."

Anahera busied herself, setting the table and unpacking containers filled with steaks, potatoes, carrots, and salad.

"Right now, you are thousands of miles away from

all of that," she says, retrieving a wine bottle from the refrigerator. "Come on now, while your dinner is hot."

Nathan, still a bit unsteady, rises slowly from the floor in front of the computer and makes his way to the table. "My God, this is a feast!" he exclaimed.

Anahera bowed her head, murmuring a quiet blessing, before pouring wine into two goblets.

"Enjoy," she said softly.

Nathan ate in silence, fully engrossed in the meal.

"Damn, Anahera, this is incredible!" he remarked between mouthfuls.

"Here at Half Moon Resort, nothing is too good for the wealthy guest," Anahera replied with a smile.

"Right," Nathan nodded, his attention fully captured by the food.

"My friends call me Ana," Anahera offered.

Nathan paused mid-bite, setting his fork down to regard Anahera seriously.

"Sorry," he said, "I was making a pig of myself. I forgot how hungry I was or the last time I had a meal like this."

"Well," Anahera smiled gently, "we know now it was over eighty years ago."

Nathan chuckled, but then his expression turned serious as he focused on Anahera.

"Anahera... Ana, why are you doing this?" he asked. "You could have turned me over to hotel security when you found me last night."

"You were hurt," Anahera replied, "and it was obvious you were frightened and in trouble."

"I could have killed you," Nathan confessed.

"No, you wouldn't have," Anahera assured him. "You were just scared and alone, lying there facing a woman."

Nathan adds, "A very attractive woman with a dangerous cell phone," and they both laugh, the tension of the moment dissipating.

He sips the wine and looks at Anahera. She realizes he is not eating but is clearly exploring her face with his eyes. She smiles. Softly, he says, "Thank you, Ana."

***

University-Maryland football team is on the field as two vehicles, each moving at a slow pace along the back fence, converge midway. The distant melodies of the school band waft through the air. Simultaneously, the passenger window of one car and the driver's window of the other slide down. Admiral Davis occupies the driver's seat, passing a metal briefcase to the handsome young man, Maxwell Lewis, in the passenger seat of the second car.

"Everything you will need is in the case, including an encrypted satellite phone," Admiral Davis informs him. "Have a nice vacation."

As Admiral Davis drives away, leaving behind the jubilant cheers of the game crowd, the second car follows suit.

An attractive young Korean woman, Konnie (pronounced "Cony"), sits in the driver's seat, skillfully maneuvering the vehicle through the bustling parking lot and out into the Maryland afternoon traffic. Meanwhile, Maxwell (Max) opens the case and retrieves two airplane tickets and a colorful brochure.

"What gifts did the Admiral give us this time?" Konnie inquires.

"Two roundtrip tickets to Half Moon Bay private resort in the South Seas," Max replies, examining the contents.

"Lovely," Konnie acknowledges.

Max continues, "Reservations for Mr. and Mrs. Hershel and Cho Bromberg."

"We depart tonight to Manila Ninoy Aquino International Airport," Max elaborates. "We will be escorted to a private hangar and board a seaplane that will take us to the island."

Max scrutinizes the brochure before urging, "Go online and check this private resort out."

He withdraws two passports from the case, offering one to Konnie. She glances at it briefly before returning it to him. Max then retrieves a bundle of cash and two black credit cards, stowing everything back into the case alongside the satellite phone. He retrieves a file from the exterior pocket of the case and flips it open.

"Who's the target?" Konnie queries.

"Anahera, a Filipino nurse who works in the resort clinic," Max reveals. "We are to observe but not approach until further notice. The Admiral has made special handling of our luggage, so make sure you include what we need; there is a list here in the case."

Chapter 5:
Two New Guests to Moon Bay

Anahera meticulously finishes stitching up Nathan's head wound and gently applies a bandage over it.

"Done," she announces softly. "That wasn't so bad. When the lidocaine wears off, it's going to hurt for a few days."

"A few days?" Nathan grimaces. "I can't stay here for..."

Anahera interrupts, her tone gentle yet firm, "You have somewhere you need to be?"

Nathan hesitates before conceding, "I guess not, but I can't stay here."

"May, June is the rainy season," Anahera explains. "The resort will not officially open again until next month. There are only a few guests here, mostly for the hot sauna treatment. The majority of the security is on vacation."

"Thank you. I am going to need to get to my ship and retrieve my sea bag, charts, and file," Nathan informs her.

"We can figure that out tomorrow. I've told the clinic

I will be taking a few days off," Anahera reassures him.

"Okay. I am also going to need some tools," Nathan adds.

Anahera glances towards the bedding she has prepared for him. "Now, I suppose the military taught you how to make a couch into a bed?"

She gestures towards the makeshift bed equipped with sheets, a pillow, and a blanket. "I am going to bed," she announces, giving him a quick kiss on the forehead before walking away. "Good night."

"Ana, thank you," Nathan calls after her as she exits the room.

"Give the computer a rest and go to sleep," Anahera's voice echoes from the distance.

Alone in the dimly lit room, Nathan retrieves a paper from his jacket pocket and his orders from Charlie before he departed. He gazes at the document, its coordinates indicating his destination. Moving to the computer, he opens the Google Earth program and enters the longitude and latitude numbers from his orders.

As the earth image turns and slowly zooms in on an island, Half Moon Bay, Nathan is left staring at the screen, contemplating his next move.

***

Admiral Davis sits at a picnic table, enjoying his morning coffee while perusing the Washington Post. He places another coffee on the table as Colin Brown arrives on a speed bike, carrying a bag.

"Morning, Jack," Colin greets, dismounting. "Brought you the Sunday New Yorker. You'll find the sports section interesting."

Admiral Davis folds back the paper to reveal the file and says, "Thanks."

"We ran another satellite pass and picked up an unknown anomaly. The coordinates are in the file." Colin Brown adds.

"How's our girl doing?" Admiral Davis questions.

"So far, she has spent a lot of time online researching world history," Colin Brown replies.

"World economy, religion, politics?"

"No, war," he adds. "From Wilson's expedition into Mexico, the First World War, Pearl Harbor, Hitler, the bomb, Korea, Vietnam, Afghanistan, Russia, China, the world recession in 22, and the recipe for buttermilk pancakes."

"Strange subject for a young woman on a resort island in the middle of the South Pacific. What else do we know about this nurse?" Admiral Davis asks.

"She was born and raised in an all-girls Catholic orphanage and, after high school, was sponsored by some billionaire, along with four other women from the orphanage, to attend the College of Medicine, Nursing and Health Sciences, Fiji," clarifies Colin.

"After graduation, did this sponsor send the five to this island?"

"No," Colin replies, "only Anahera."

"Interesting. As you said, it is probably nothing."

***

Nathan stands by the window, dressed in his clean and pressed flying suit, observing through binoculars. He sees a covered, well-styled, six-seat golf cart ascend from the beach. The driver and two women occupy the first guest seats while Max and Konnie sit behind them. Another cart, different in style, trails behind with luggage.

"We have visitors," Nathan announces.

Anahera enters from the bedroom, brushing her hair

and clad in khaki pants and a colorful island blouse.

"That would be Mrs. Mildred Wentworth and her maid Tilly. They come twice a year for the hot mud bath and spa," Anahera explains.

"Also, a couple and a lot of luggage," Nathan observes.

"The luggage is probably Mildred's. The couple are..."

A pause.

"Hershel and Cho Bromberg. Fishing and snorkeling. First-time guests," Anahera finishes, checking her cell phone.

"This time of year?" Nathan questions.

"Spotted Sea Trout, Mangrove Snapper, Sheepshead, Redfish," Anahera replies.

Nathan hands Anahera the binoculars, and she places them in a picnic basket.

"Are you ready to go?" she asks.

"Do you have the tools?" Nathan inquires.

"Already in the beast out back," Anahera replies.

They exit the cottage to find a Coleman Outfitter

electric 400 UTV parked at the edge of the jungle.

"The beast?" Nathan questions as he settles into the passenger seat.

"It eats jungles. Belt up," Anahera urges as they embark onto a narrow vine-leafed path leading into a canopy of trees and large-leafed jungle plants. "There is only one path from the north beach that could have brought you to the cottage."

***

Max, clad in scuba shorts and a wet suit vest, finishes setting up a scope on a tripod aimed at the cottage. Konnie enters.

"I have reserved a boat and picked up some snorkeling gear and fins," Konnie informs him.

Max nods, adjusting his focus. "The Admiral called. He has something just off the beach on the north side of the island for us to check out and then call him."

She hands him a beach umbrella in a long cloth bag and a wicker picnic basket.

"Take the umbrella out and put the M-16s in the bag and the Berettas in the basket with the sandwiches and champagne," Konnie instructs.

Max complies, picking up the weapons and begins to place them in the umbrella.

"They had photos of the hotel staff in the lobby and I got a shot of the women," Konnie mentions casually. She then slips out of her dress, dons a small black bathing suit and a tight rubber vest, and retrieves a black kimono from the closet.

Max raises an eyebrow at her attire. "What time is the boat reservation?"

Recognizing the look. "Later, darling, business first," Konnie replies with a wink, emphasizing their priorities.

Chapter 6:
More Snakes In The Jungle

The beast sits near the bright blue Hellcat, its fuselage engulfed in jungle growth. Nathan removes the leather tool bag from the beast and places it at the lower edge of the starboard wing. Nearby, Anahera pulls some of the palms from the aft fuselage and the tail section.

With a machete in hand, Nathan hacks away at the vines covering the wing area leading to the cockpit. Standing on the wing, he tries to slide the canopy open. He takes the hand grip of the machete and pounds on the lower edge of the canopy, then the top edge. As the slide finally gives way and the canopy opens, Nathan finds himself staring right into the eyes of a Samar cobra with its golden head popping up from the cockpit, ready to strike. Nathan slams the canopy shut in time to catch the cobra's neck between the canopy and the windscreen. He holds the canopy tight and, in one movement, brings the machete down on the cobra's neck, severing it from its body into the cockpit. In that motion, he takes the skin off the side of his wrist as blood runs down his arm. He drops the machete and grabs his wrist.

Anahera quickly climbs up on the wing, takes his hand, and applies pressure to his wrist, stopping the bleeding.

“Did he bite you?” she asks urgently.

“No, but the machete did,” Nathan replies.

Anahera takes the white bandana from around her neck and wraps Nathan's wrist with it. She pauses, then checks to see if the bleeding is serious. She turns his wrist over as though she is looking for a cut.

“It's healed,” she says in amazement.

“What?” Nathan looks down at his wrist.

“That's rather strange. Look at me,” Anahera instructs.

She reaches up and lifts the edge of the bandage on his forehead. Looking at the wound area, she takes the bandage completely off. The sutures are gone, and only a slight scar remains.

“Remarkable,” she mutters.

“What is wrong?” Nathan asks, a hint of worry in his voice.

“Your body is healing at an accelerated rate,”

Anahera explains.

"Will I be alright?" he asks.

"Apparently," she replies.

Nathan reaches behind the Hellcat's headrest and pulls out his sea bag.

"Good," he says, handing her the sea bag. "Empty this in the trash bag you brought. I need to put something else in it."

She jumps down and goes to the beast as Nathan moves under the fuselage to drag the tool bag with him. Under the plane, Nathan takes a large Phillips head screwdriver from the bag. Anahera kneels down, looking under the fuselage.

"You don't need that. Try this," she says, reaching into the bag and handing him a battery-powered drill with a Phillips head bit.

Nathan looks at it and pulls the trigger on the drill. "Nice. Do you use this in the OR?"

Anahera smiles and moves out, going back to the beast. She continues to take the folded clothes out of the sea bag and puts them in the black plastic garbage bag.

"You know, I was thinking. If you go back, they will

either treat you like a hero or a lab rat," she says.

Nathan nods, "It has crossed my mind."

Nathan removes the first screw head and continues taking the screws out of the round shield that holds the white soup bowl-locking cover in place. Setting the shield aside, he begins to remove another set of screws.

"Bring me the sea bag," he calls to Anahera.

Anahera brings the empty sea bag to Nathan, who is climbing out from under the plane. He holds the device shaped and sized like a garbage dispenser. She holds the bag open, and he puts the device into the sea bag with some other parts, including the plastic soup bowl cover. As he ties the bag closed, a blue glow emanates from within the bag at the top of the device. Nathan then carries the bag to the beast.

Anahera hears something and walks out of the jungle cover to the edge of a large leaf plant, looking out at the ocean.

"What is it?" Nathan asks from the beast.

Anahera's eyes focus on a small boat approaching about a quarter of a mile south of their covered area.

"A small boat," she says, moving back further under the large plant.

Nathan hurriedly throws some large palms over the tail section and the starboard wing of the plane. “A fishing boat?” he asks.

“No, I don’t think so,” Anahera replies. “They're just moving slowly north, and one of them has binoculars looking at the shoreline.”

“Is it the hotel security?” Nathan places the tool bag in the beast.

“No,” Anahera responds, backing up slowly and joining Nathan at the beast within the jungle cover.

Aboard the Sportsman Heritage 251, Konnie is at the center console, and Max scouts the shoreline through the binoculars.

“I think I found her. Slow down, Konnie. She just moved back into the shadows,” Max says. He places the satellite phone to his ear, touching a series of numbers.

“Yes,” Admiral Davis answers from off-screen.

Max responds, “We found her, and I think your Grumman F6F Hellcat.”

“Is she alone?” Admiral Davis asks.

“Not sure yet. Hold on,” Max says. He takes the phone from his ear and leans toward the island, yelling,

"Hello there, on shore. We are from the hotel. Is Nurse Anahera there?"

Nathan steps out of the shadows onto the beach, wearing his shoulder holster. He stops at the edge of the jungle and looks in the direction of the boat.

"What can I do for you?" he yells back.

Max, holding the phone, looks through the binoculars as the boat rocks, with Konnie trying to keep it in place.

"We're from the hotel," Max yells. "They have an emergency and sent us to find nurse Anahera." Lowering his voice, he speaks into the phone, "Hold on, I think we made contact. It may be him, the pilot."

Admiral Davis's voice crackles through the phone, "Are you sure?"

"Yes," Max replies.

"Unbelievable!" Admiral Davis exclaims.

On the shore, Anahera's voice is calm but firm, "He's lying. I have an emergency signal in my cell phone."

Nathan echoes, "She's calling the clinic right now."

Max, aboard the boat, calls out, "Do you mind if we

come ashore? We are actually here to help you home, Lieutenant." Lowering his voice, he speaks into the phone, "It's him!"

"If you don't think you can take him alive, kill them both. I want the device on that plane!" Admiral Davis orders.

Nathan's voice is filled with defiance. "Go to hell!"

The boat's twin 200 HP Yamaha V-6 outboards roar to life as Konnie turns it inbound. Max raises an M16, trying to aim as the boat bounces on the waves.

Nathan's hand moves quickly, pulling the Remington M1911A1 pistol from his holster. He aims and fires two rounds.

The bullets shatter the center console windshield, sending plastic splinters back at Konnie, hitting her face and arms. The boat makes a sharp turn as Max sprays the beach with bullets. Nathan runs for the beast.

Anahera has the beast ready as Nathan leaps into the passenger seat. She turns and heads into the deep green canopy. Off-screen, Max's weapon continues to blast a path along the jungle's shoreline.

"Who the hell are they?! And how did they know I was here in the middle of nowhere?!" Nathan demands.

Working to control the beast, Anahera plows through the thick jungle foliage, traveling deeper and further north. "My guess? Someone hacked my computer when we entered your name."

"Hacked?" Nathan echoes, bewildered. The sounds of Max's continued firing echoes through the jungle.

Anahera explains, "Let me put it this way: when anyone puts anything online, someone somewhere will read it."

"That doesn't seem right," Nathan grumbles.

"I would have to guess that some government agency has a watch list for downed pilots," Anahera speculates.

"There is no way we can go back to the hotel," Nathan asserts.

They are jostled around as the beast speeds through the jungle. "I may know a place where we can be safe long enough to figure out what we do next," Anahera suggests.

"We?" Nathan looks at her.

"Yes," Anahera confirms.

Nathan holds onto the top cover of the beast. "Ana, I am sorry."

There is a beat of silence before Anahera speaks again. "You know, Nathan…"

He interrupts her, "My friends call me Nat."

They hit another bump, and his head bangs against the top.

Anahera continues, "Growing up in an all-girls orphanage, not knowing who I was or where I came from, my life was far from eventful. College in another country, my life has been pretty normal. Boring is a better word."

Lt. Nathan Miller grins wryly. "Until you found a man lying on your living room floor and discovered he was born in 1922."

Anahera laughs. "Yes, but let's not forget a man who can send antibodies, white blood cells, growth factors, enzymes, and nutrients to the injury in seconds and close a wound in hours."

Max's weapon continues to fire blasts from the distant shoreline. "And have you been involved in a shootout before lunch?" Nathan adds.

"To be honest, Nat... I am having the time of my life. And now I know how to make whipped buttermilk pancakes," Anahera responds.

They hit another bump, both colliding with the beast's roof. There’s a beat of silence.

“Where are you taking us?” Nathan asks.

“There are some sinkholes and caves at the far northern end of the island, and a buried building” Anahera explains.

***

Max, dressed in island pants and a shirt, is on the satellite phone with Admiral Davis. Konnie lies on the bed; her right chest and arm are bandaged, and her face is covered with a bandage resembling the “Phantom of the Opera” mask.

“I think the hotel bought the bird strike story,” Max says.

“There was no sign of the device on the aircraft?” Admiral Davis inquires.

“Nothing but a round hole in the fuselage,” Max reports.

“Max, they cannot get off that island. Do you understand? You must get that device at any cost,” Admiral Davis's voice is stern.

“Yes. After what happened, I don’t think they will

come back here or contact the Navy or the government," Max assures him.

"All these years, I have been looking in the wrong direction...out of Leyte. I was sure Monnada was where the 'Cold Blue' operation was located," Admiral Davis reflects. "It's there on that island! Now I am sure of it."

"Best we could tell, they headed north into the jungle," Max updates.

"I've been looking for a Navy requisition order for the Seabees to construct a special type of building in the Philippines around 1941/42," Admiral Davis reveals.

"There must have been hundreds built during the war," Max reasons.

"Not to the type of specifications this building called for. I located a work order yesterday given to the Seabees on June 20th, 1942, which would eliminate a runway and bury, not destroy, a steel building. The building would be buried under rocks and dirt and completed before the Japanese took the island on June 30th, escaping the Marine landing. In July, the Japanese lost the island to the United States. Where was this island?" Admiral Davis explains.

"This island?" Max's interest piques.

"Yes. I am sending you the coordinates where the building was buried. Check it out, and again, you must get that device. If you can get in the building, pack up anything of note. If that is where they have headed for, find and kill them. Max, finish this and clean it up!" Admiral Davis orders.

"Yes, Sir," Max affirms.

The Admiral ends the call. Konnie's eyes burn with a vengeance. "I will find the son of a bitch, cut him to pieces, and before he bleeds to death, put a bullet in his ear."

Max remains composed. "I reserved a four-wheel, electric terrain vehicle. If you can make it, we leave at sunset. We should have the Admiral's coordinates by then."

Chapter 7: The Half Moon Bay Island Laboratory

A cave is partly covered by a small waterfall that forms a pool at its base. The beast sits just inside the narrow cave opening. Nathan follows Anahera, her flashlight guiding their way deeper into the black, walled cave. They approach an area where the wall has crumbled, revealing a narrow path off to the right.

They step over the rocks and loose dirt rubble and enter a small room with a wide steel door.

“See if you can pull it open. I haven't been up here in a while,” Anahera says.

Nathan sets down the sea bag and grips the steel horizontal bar, giving it a pull. The door resists but then begins to open.

“What is this place?” Nathan asks.

“When I first came to the island, I discovered it and used to come up here to study,” Anahera explains.

As they open the door, several colorful birds scatter towards an opening in the ceiling and disappear.

A blue glow escapes from somewhere inside the open door. They step through the steel door into the room. Anahera's flashlight moves slowly about the room and holds onto a dust-covered large logo of a blue ten-point star attached to the wall with the letters: “COLD BLUE”. Illustrations of lightning surround the letters.

“A flash of lighting and a cold blue light is what I experienced before passing out,” Nathan says, staring at the logo.

“This morning, I checked online the latitude and longitude coordinates of this island against my ordered destination. I made it but a few decades late.”

Anahera holds the light on the logo. “The ten-pointed star is an upside-down pentagram on top of a right-side-up pentagram, which is the symbol of man as a star representing law and order in the realm of physics.”

Nathan looks at her inquisitively. Anahera shrugs, “Catholic school.”

Nathan starts to move towards two dusty glass doors.

“Instead of arriving here June 17th, 1944, somehow

I was transported from my location and crashed here June 17th, 2025," Nathan mutters.

He opens the double glass doors leading to the other room and enters, Anahera following with the flashlight. The room is large and mostly filled with dusty, waist-high tables and stools. In the center of the room is a glass dome over a metallic round box. A soft blue light travels upwards, filling the dome and lighting the room.

"The blue light?" Nathan asks.

"It was there and hasn't been deemed or changed," Anahera responds.

Looking into the dome, Nathan responds, "Somehow, that light is traveling up from far below the building."

"The room I studied in is through that door," Anahera says, leading Nathan to another room off this obvious work area.

The room is a large office with a dozen partly emptied bookcases, two plastic-covered leather couches, a large desk, and a glass-enclosed square workstation. A single ten-pointed star about the size of a basketball is mounted inside a clear Plexiglas ball with some form of locking device. It is sitting on the desk, illuminating the room.

"Where is it getting power?" Nathan wonders aloud.

"Never gave it any thought," Anahera admits.

Nathan walks to the large, faded, somewhat ragged map on the wall behind the desk. The logo is at the bottom, along with the words "COLD BLUE". In script letters, it reads "MONNODA ISLAND; Commandant: Commander Brereton Lewis, USN."

"Commander Lewis. The officer I was ordered to report to," Nathan murmurs.

The island's shape is distinguished by the large shape of a cobra's head at the north end and a half-moon bay at the south end. There is a sizable building depicted at the far north end in the eye of the cobra. There is also an adjoining airstrip depicted.

Anahera says, "Through those doors is a kitchen and an empty warehouse that I believe had an elevator at one time. It also has what I believe is a glass-blowing Furness."

Nathan opens the drawers of the desk and the file cabinets. A few odd-typed sheets of paper and notes remain, but nothing else. He looks down at the dusty desktop. There is a coffee cup and a carved pipe sitting in an ashtray. Nathan blows the dust off a desk calendar.

The Date Is June 18, 1944. Nathan looks around the room as Anahera moves to a round table with crystals in various configurations.

"What the hell happened here?! It's as though they just packed up everything and left in a hurry," Nathan exclaims.

Anahera picks up one of the small glowing crystal spikes and puts it down. "It's warm. Question is, what did they design and manufacture here?"

"Obviously, what's in my sea bag," Nathan replies.

"A teleportation device...and..."

"A light that needs no external power."

Anahera pulls the plastic sheets off the two couches. "Make yourself comfortable. I don't think your friends are going to be stopping by tonight."

Nathan continues walking around the room. "Who sent them? My government or someone looking for all this?"

"Whoever they are, I don't think they want to bring you home for a parade down Pennsylvania Avenue," Anahera says.

Nathan sits in the chair behind the desk and studies the map on the wall. "I'm going out to the beast before the sun goes down to get the picnic basket and tools," Anahera announces.

"Can you also bring in the bag of clothes?" Nathan asks as Anahera exits.

Nathan continues to study the map, his eyes moving to the right-hand side. He notices a crack along the edge of the frame, emitting a blue glow. Nathan stands and moves close to the right side of the frame, noticing the blue light filling the crack. He touches the frame and presses on it. Something is illuminating the area behind the framed map, but pressing on the map itself does nothing. Running his finger along the side and bottom of the frame, the wall behind the frame suddenly slides to the left, exposing a room dimly lit by various sizes, blue and glowing ten-star crystals in clear empty containers positioned about the room. Also scattered on a large rubber mat on a metal table are additional, versus size, glowing crystal spikes.

Nathan steps into the room, the soft hum of the crystals filling the space. He walks over to the table, marveling at the array of crystals, each one pulsing with its own rhythm of light. The blue glow creates an otherworldly atmosphere, casting long shadows and shimmering reflections on the walls. Also on the table are several sets of rubber gloves.

Anahera returns with the picnic basket and tools, pausing as she sees the newly revealed room. "What did you find?"

Nathan turns to her, the blue light dancing in his eyes. "A lot more than just a teleportation device."

They stand together, absorbing the scene. "This place," Anahera says softly, "it's like something out of a dream. Or a nightmare."

Nathan nods. "Whatever happened here, it's not over. I think we need to determine the purpose of this place and the relationship between these crystals and the device that transported me to this island. We need to find answers before whoever is out there finds this building and us."

Anahera sets down the basket and tools and steps closer to the table, examining the crystals. "They must have been working on something very secret, something beyond our understanding."

"Yes, light without lead wires or a generator." He states the obvious. "The power that operated the device."

Nathan looks back at the map, the cold-blue logo, and the details of the island. “Why was I sent here to meet with Commander Lewis and his team? Was I the experiment or just a delivery source?”

They sit down on the couches, pulling out the sandwiches from the wicker picnic basket. Their eyes keep drifting back to the room of crystals, the soft blue light a constant reminder of the mysteries they still need to unravel.

“Do you think anyone else knows about this place?” Nathan asks between bites.

“I doubt it. Maybe a few islanders,” Anahera replies. “But we have to stay vigilant. Whoever sent those people will keep looking.”

Anahera replies, “That’s a good question, but first, let’s eat. If you’re going to be at this all night, you need your strength.”

“Yeah, while we can. The only thing I can guess is that I was carrying a working device on my aircraft, and it was to be handed over to Lewis. I can’t explain why else they would pull me from a major air-sea operation to bring this thing to them. My ship and that device had to be a part of all this. Apparently, the teleportation device transported me to the island, but not to 1944—rather, to 2025. It is the right island, but the wrong time. I think when I did not arrive, the project was scrapped, and they were ordered to clear it."

Nathan nods. "We'll figure it out. Together."

Anahera smiles. "Together."

Nathan looks at the glowing crystal spikes and begins to evaluate the question. "The shaft in the warehouse is obviously what they used to bring up the crystals from some form of mine below. There must be magma or lava caves there, which, over time, influenced the various unique minerals in the surrounding ocean waters. Most of these islands were formed by volcanoes, and as the magma continued to cool, quartz crystals began to grow as more silicon and oxygen combined with an unknown element. What emerged were these spikes. The question that remains is, what gave these crystals unlimited power?"

"Everything in the universe, including humans, apart from energy, is made of matter. Therefore, atoms make up everything in the universe. We know atoms are made up of three particles known as subatomic particles: protons, neutrons, and electrons. What part of all this gave the device the ability to alter the shape, contours, and contents of all matter and transport that matter without destroying the billions and billions of atoms that make up the very matter of the subject element? Impossible, but here I am."

Anahera makes a point. "Something within what you have assumed altered the composition of your white blood cells, increasing the full range of accelerated healing. We have no idea what else the event altered in your internal makeup."

Nathan smiles, and the seriousness of the moment lessens. "It will be interesting to discover exactly what else. It could be all good."

Anahera moves to the couch next to him, and Nathan puts his arm around her. She looks up at him and whispers, "As we agreed, together we'll figure this out."

As the night deepens, they plan their next steps, the glow of the crystals casting a hopeful light on their path forward.

Chapter 8:
A Double Eagle

While Anahera decides to explore the other rooms, Nathan steps into the hidden room, and the wall closes behind him. He walks closer to the table and leans forward to look at the hundreds of blue-glowing, needle-like crystals and the various clear Plexiglas balls holding the ten-pointed stars. It's clear that a departure was made in a rush. Nathan finds a ledger on a small table in the back of the room. He sits down, opens it, and begins to turn the pages. He looks up as Anahera's voice breaks his concentration.

"Nat... Nat?" Anahera calls from outside.

Nathan gets up and opens the panel. "In here," he responds.

Anahera sets the basket and clothes bag down as she enters the room. "They are beautiful," she says, looking at the crystals.

Nathan sits down and returns to the ledger. Anahera walks about the room, admiring the glowing crystals.

"The professor sealed this room off but left behind

his day calendar," Nathan says.

"And these crystals in various stages of assembly, they are warm," Anahera adds.

Nathan explains, reading from the ledger, "In June, he was ordered that Washington was about to close down this operation."

While Nathan is talking, Anahera continues to look around the room. Random sketches and engineering drawings are pinned to the wall, and a few are scattered on the floor. She picks them up and places them on the table. "Apparently, the last three teleportation demonstrations had failed and resulted in a number of deaths. The Hellcat experiment was their last chance to convince Washington to keep the project going," Nathan continues.

"You showed up on the island as planned, but eighty years later," Anahera says.

"On June 20th, the full force of the Seabees arrived, and they were given twenty-four hours to vacate the building," Nathan recounts from the words in the ledger.

"Why bury it? Why not just demolish it?" Anahera asks.

"They destroyed the airstrip, but the professor and

Commander Lewis convinced Admiral Turner, Navy Intelligence, that they had found something extraordinary on the island," Nathan explains.

"The crystals, 'Cold Blue,'" Anahera realizes.

"Yes, the decision came down to bury it and return after the war that was already turning in favor of the Allies," Nathan says. However, something else happened, and no one returned.

"They moved everyone and everything out in twenty-four hours, hence the mess left behind," Anahera concludes.

Nathan adds, "Whatever secrets they found, the truth is sealed in this room and buried for 80 years. I am now sure the crystals, however they were formed, are some kind of continuous energy source."

"Based on what we discovered online, they all died before the bomb ended the war a little over a year later," Anahera notes.

"The question is, who is now trying to close the book on the 'Cold Blue' program by eliminating a loose end, us?" Nathan wonders aloud.

"That is one way to put it. Nat, if they believe you're

actually here and this is some government-classified secret, they may be out to silence both of us," Anahera warns.

"What did Naval Intelligence do with all the equipment that was taken out of here?" Nathan asks, his mind racing.

Nathan and Anahera continue to sift through the remnants of the past—the charts, the ledger, and the Plexiglas balls holding the ten-pointed stars—each piece of the puzzle bringing them closer to the secrets of "Cold Blue" as the threat of their pursuers looms ever closer.

***

The Admiral is talking on his desk speakerphone to Colin Brown.

"I haven't discovered anything more. I will be sending an agent to the island Wednesday," Admiral Davis says.

"Do you remember talking to a retired Colonel David Winnkemp?" Colin Brown asks over the phone.

The Admiral writes the name on a yellow pad. "Colonel David Winnkemp... not that I can remember. I must have spoken to over a dozen officers in the Philippines from 2000 to 2002. Why?"

"He gave the order to move an office and warehouses in June 1944, at the coordinates I gave you on Sunday," Colin Brown explains.

"Where did you find that?" Admiral Davis asks.

"According to the Navy's Seabee archives, before he died in 2000, he wrote a book called 'BATTLE READY EARTH MOVERS,'" Colin Brown replies.

"I will have to check that out," Admiral Davis says, intrigued.

"I just emailed you a copy. Winnkemp gives a vivid description of a building's contents. He moved from an island and then ordered the building to be buried," Colin Brown informs.

Admiral Davis leans back in his chair, contemplating the new information. "This might be the lead we need."

Admiral Davis leans back in his chair, a thoughtful look on his face. "Could be our island."

Colin Brown's voice crackles over the phone. "He also states that he had to get some professors and some Navy personnel out of there overnight."

"Interesting," Admiral Davis muses.

"Now, here's the best part," Colin continues. "Along

with boxes of files, they removed several packets containing five million dollars in American greenbacks and three bags of the Liberty twenty-dollar gold coins saved from the Mint meltdown of 1937. As well as sealed container holding tubed shaped gold ingots."

"Gold!?" Admiral Davis sits up straighter.

"Yes, it seems they were paying the locals in gold coins who refused to take paper money," Colin explains.

"Did Colonel Winnkemp mention what happened to the equipment, files, and money?" Admiral Davis asks, his curiosity piqued.

"No, but he says in the book he was contacted by someone from Naval Intelligence," Colin replies.

"I don't remember finding any reference to a Colonel Winnkemp in our files," Admiral Davis says, frustration creeping into his voice. "I will damn well get someone on it in the morning."

"Probably a good idea, Jack. Good night," Colin concludes.

The Admiral sits for a moment, then opens the drawer of his desk and takes out a twenty-dollar Liberty gold coin mounted in plastic as a paperweight. He examines it thoughtfully.

***

Nathan is still sitting at the small table, reading the ledger. Anahera has taken the device out of the sea bag and places it in a canister sitting on one of the tables. The top is on some sort of hinge, so she swings the top in place.

"I think I found the carrying case for the...whatever it is," Anahera says.

Nathan looks up to see what Anahera is talking about. "Great! Did you say there was something more in the basket to eat?"

Anahera brushes her hands off and looks around for a sink or bathroom. "I know there was a bathroom somewhere here."

A door at the back of the room is partly open. She moves the door further open, steps in, and calls for Nathan. "Nate... Nate!"

Nathan looks up and turns to see where her voice is coming from.

"I think you better come and look at this," Anahera says.

He puts down the ledger, walks to the open door at

the back of the room, and steps in.

They look at each other, silhouetted by the soft blue light, and then at the shelves in the room placed above an open, emptied safe.

One of the four shelves holds a package, which is about thirteen inches by thirty inches wide and twelve inches high.

"Are they real?" Anahera asks.

Nathan touches the square sticker on the packet and reads the label out loud. "Bureau of Engraving and Printing, Washington DC, United States of America. Secretary of the Treasury - Henry Morgenthau, Jr., Franklin Delano Roosevelt, President of the United States of America. Looks like this is the real thing."

On a lower shelf are two canvas bags. One is partly opened, and some of the contents have fallen out onto the shelf. Anahera picks up one of the coins with an eagle picture on it.

"A rather pretty coin," she comments.

Nathan looks at the coin. "Turn it over."

Anahera turns it over and hands it to Nathan. "This is a twenty-dollar, 1933 double eagle gold coin," he explains.

"Gold?" Anahera asks.

"Yes," Nathan confirms. "On Leyte, paper money was worthless to some islanders, so the Navy was paying Philippine officers and high-end laborers in gold and others with specially marked money. The professor must have been doing the same here."

They both just stand there looking at the money for a beat, somewhat in a daze. Nathan notices a metal box about the size of a cash box on the top shelf, brings it down, and opens it. Anahera looks at the four tube-shaped objects about the size of a fountain pen and about a foot long on the inside of the box. "What are they?" she asks.

Nathan picks up one item and takes the wrapper off of one. "Gold rods of some kind."

Nathan stares at the money, coins, and gold rod he holds, his expression troubled. "We can't just leave all this here," he murmurs.

Anahera nods slowly, her own gaze fixed on the package of bills. "Yes... I mean, no, we can't."

"Right," Nathan agrees, a sense of urgency creeping into his tone.

Anahera breaks the silence, her voice gentle. "Did

you say you were hungry?"

Nathan's stomach growls in response. "Yes."

"I'll go fix us something," Anahera offers, turning to leave.

"I'll put all this in my sea bag," Nathan decides, determination firming his resolve.

"Okay," Anahera replies softly as she exits the room.

Nathan gathers up the canvas bags and carries them over to his sea bag, where Anahera had previously removed the device. Carefully, he places the coin bags inside, then turns his attention to the scattered money and the box of gold rods.

Chapter 10:
Little Uninvited Creatures

Max enters the dimly lit room, finding Konnie still wrapped in bandages, diligently packing her belongings.

"I have checked us out," Max informs her, his voice echoing in the quiet room. "Told them we will be leaving in the morning."

Konnie looks up from her task, concern etched on her face. "What about getting off this damn island?"

Max begins to pack his own clothes, his movements efficient. "I told them we want to get you to your doctor in Mindanao, and we have arranged for our own transportation before noon."

"Did they object?" Konnie inquires, her tone wary.

"No," Max replies, a hint of reassurance in his voice. "I gave them the name of the company. It's one they have used before."

Konnie finishes closing her suitcase, a sense of unease lingering in the air. "Will we be coming back here?"

Max closes his own suitcase with a decisive click.

“No. We'll head out before sun up, do what needs to be done, and have the steamer pick us up on the north beach pear.”

“I'm not thrilled about going through that jungle before sun-up,” Konnie admits, a shiver of apprehension running through her.

In the stillness of the room, their preparations complete, Max and Konnie exchange a look, silently steeling themselves for the challenges ahead.

***

Nathan lies on one of the two couches, a sheet draped over him. He's changed into his military khakis and a white t-shirt. Anahera, now wearing a gray, navy t-shirt with her long blouse tied as a skirt, sits in the blue light on the second couch, watching Nathan sleep. Slowly, she gets up, wraps part of the sheet around her arm and right hand, and approaches Nathan. She retrieves his USN K2 knife from the picnic basket, places it in her sheet-covered hand, and moves cautiously next to the couch.

“NAT, don't move,” she whispers.

Nathan's eyes open, blinking in confusion. “What?”

“Don't move,” Anahera repeats.

As the words leave her lips, the knife plunges

downward next to Nathan's face, driving the blade into a black widow spider. With her other hand, Anahera swiftly pulls the sheet with two other spiders off Nathan and onto the floor.

Nathan jumps up, now fully alert, as Anahera kicks another spider away from his bare foot and crushes it with her shoe-covered foot. She sets the knife down gently and looks up at Nathan, who is scanning the floor and couch. Their eyes meet for a moment, and they embrace. Nathan holds her tightly, feeling her body pressed close to his. She slowly steps away, and they remain silent.

Anahera glances down and, in one swift move, knocks Nathan back onto the couch and steps on the third spider. Nathan looks at the crushed black widow and then up at Anahera.

"You get used to them after a while," she says with a shrug and a smile.

Nathan just shakes his head. "You're incredible..." he says and reaches for her, "I..."

"It's almost sunrise," Anahera interrupts. "Shouldn't we be doing something about, you know, those people?"

"We need to move the beast," Nathan says, putting

on his socks.

"I did that when I got the basket and clothes," Anahera replies, slipping on her khaki pants.

"Chances are they have no idea this place exists," Nathan speculates.

"I don't suppose there was any secret about this place. As I said, the island people may know of it, but when I came here, no one was ever around," Anahera counters.

Nathan takes a dark green shirt from his pile of clothes and hands it to her. As he slips on his khaki shirt and puts the shoulder holster on, he puts three mags in his pocket. Anahera picks up the MK2 knife, places it in its sheath, and ties it around her waist.

"If they do show up, I don't suppose it would do any good to tell you to hide in the safe room," Nathan suggests.

"No," Anahera replies. "They're going to find their way in here, and I know where we can get a better view of who comes through that door."

"What do you have in mind, Captain?" Nathan asks as she walks over, picks up the flashlight, and heads toward the lobby. He follows her.

"In the lobby, there is an attic where the birds fly up

to. If we can get up there, we can be over them when they enter," Anahera explains.

They enter the lobby and look up at the square hole in the ceiling.

"More spiders and snakes?" Nathan jokes.

"Food for the island birds. It should be okay. Will this work?" Anahera asks.

"I think so," Nathan replies.

He looks around and notices a chair behind the reception desk. He brings the chair over the hole and turns back toward the office. "I'm going to put the sea bag up there for now," he says. We need to move the globs into the safe room, gather up the crystal spikes, and put them in the wicker basket.

"Then we wait," Anahera responds.

Anahera enters the workroom after taking the food out of the basket, grabs rubber gloves, and uses one of Nathan's t-shirts to scoop up crystals from two tables, dropping them into the basket. She pauses to place one of the smaller clear balls with a ten-star crystal carefully into the basket. And takes them to the safe room and places the basket in the safe.

***

As the first light of sunrise filters through the dense jungle, Max and Konnie arrive at the mountainside, driving their four-wheel Gator. Their luggage is piled in the back, and the umbrella bag is wedged between them.

“I don't see any buildings,” Konnie remarks, scanning the terrain.

“He could be anywhere, so keep scanning the hillside,” Max responds, his eyes fixed on the rugged landscape.

“There is a cave on the right just ahead,” Konnie points out.

“That may be it,” Max agrees.

***

Meanwhile Nathan and Anahera are in the office area. Nathan is engrossed in reading the old ledger while Anahera gathers what food that was left from what was taken from the picnic basket.

“You want a bite of something?” Anahera asks, glancing at Nathan.

“Did you hear that?” Nathan responds, suddenly alert.

"What?" Anahera asks, confused.

"Listen," Nathan insists.

Anahera pauses, listening intently. "They're here," she realizes.

Nathan stands up swiftly, urgency in his movements. "Move!" he commands.

They both rush to the lobby.

"Okay. Give me your shirt," Nathan says quickly.

Anahera hands him the shirt without hesitation.

"I'm going to climb up, and you're going to push the chair away. I'll lower the shirt, grab it, and I'll pull you up," Nathan instructs.

He climbs onto the chair, reaches for the sides of the hole in the ceiling, and pulls himself up, sliding into the attic. Anahera throws the flashlight at him and moves the chair to the wall as he lowers the shirt. She looks up at Nathan, who is peering down at her.

"Grab the body of the shirt," he tells her.

She tries once, unable to reach it. Stepping back, she moves forward and jumps, grabbing the shirt. Nathan pulls her up, helping her climb through the hole.

"You alright?" Nathan asks, concern in his voice.

"Yes," Anahera replies, catching her breath.

Nathan shines the flashlight around the empty attic area, which only covers the lobby. In several places, dirt and rocks have collapsed onto the attic ceiling, scattering debris. He shines the light upwards, finding a small opening in the sky as night changes to the purple hues of dawn.

"Well, now we sit back and see if they find the entrance. If they do, maybe we'll learn why they are here," Nathan says.

Anahera moves closer to Nathan, and he puts his arm around her.

Chapter 11:
A Crystal Blue Grave Site

Max guides the Gator toward the cave's entrance, stopping just short of a small stream of water trickling down the hillside. They get out of the Gator. Max holds his M-16, and Konnie grips her Beretta in one hand and a flashlight in the other. They step into the cave, working their way to the steel door. Max takes out a satellite phone and enters a code.

"Did you find it?" Admiral Davis's voice crackles through the phone.

"I think so. There is an entrance into the mountainside," Max replies.

"That must be it! The building was buried by the Sea-bees," Admiral Davis confirms.

"We found their Gator nearby, so chances are they're in there. Hold on," Max says.

He lifts the locking bar and slowly opens the heavy steel door.

From the entrance, Konnie shines the flashlight into the empty room as Max gazes at the soft blue light illuminating the circled logo on the wall.

"This is it...a lobby with a logo on the wall, 'COLD BLUE'," Max reports.

"Find them and that device and end this!" Admiral Davis commands. Get some photos of what is there.

"Yes, Sir," Max responds. Konnie sets the flash light down and takes out a small camera and takes photos of the logo.

Admiral Davis's voice crackles over the phone. "Make sure you take any files or anything else you may find. Set your charges to implode and get off the island." After I have the device examined, I will send in a team in to explore what was originally found in the mine. The phone clicks off.

In the attic, Anahera looks at Nathan, who holds her tight and motions that everything is alright.

***

Max looks toward the double-frosted glass doors, raises the M-16, and fires several rounds, splintering the glass into the workroom. Gravel and dirt sprinkle down from the bulging ceiling.

Konnie glances at Max with sarcasm. "Well, they probably know we're here now."

They approach the smashed double doors and step cautiously over the broken glass, entering the workroom. They walk towards the warehouse door. Konnie continues to take random photos. Max kicks it open and moves in with his weapon ready. A startled flock of birds flies up toward the raftered ceiling, surprising Max. He opens fire into the dusty, empty warehouse. What was once an elevator shaft is on the far wall, and Max walks to the edge, looks down, and again opens fire into the dark shaft.

Konnie steps out of the wheelhouse area as more dirt and gravel fall from the ceiling. She turns away and walks by the now empty work tables, where her flashlight finds the office door. Max, a few feet behind her, watches as she opens the door. He quickly moves to the side with his M-16 ready while Konnie swings around and enters the room, firing in several directions. A bullet smashes a bottle of wine, splattering glass and wine across the desk, and bullets place a dozen holes in the couch. Max enters, pointing his weapon at a pile of Nathan's clothes on the table. Konnie takes several photos and then picked up a half cup of coffee sitting on the floor next to the couch.

"It's still tepid," Konnie notes.

"They're here," Max says.

Pointing his weapon at the door on the far side of the room, he fires it, tearing it to pieces until it falls off its hinges, revealing a six-stall bathroom. The blast opens a slice in the ceiling, and gravel falls on the door.

Konnie walks over the pieces of the door into the bathroom. “You keep that up, and the mountain is going to be joining us down here,” she quips.

Ignoring her, Max slips another magazine into the M-16, turns, and looks for his next target. He looks back towards the empty wheelhouse and returns to the main room.

Konnie walks out of the bathroom and joins Max. “We're missing something. I can feel they are in here.”

“The Admiral said they originally were mining for something. The question is, what and where?” Konnie continues.

The light on the M-16 scans the cement floor, and a half-dozen spikes lie under the table. “These blue crystals. Go to the Gator and bring in the charges. We'll blast them out. Bring me something I can put these crystals in and anything else we find for the Admiral,” Max orders.

Konnie exits as Max walks through the rooms back into the lobby.

Meanwhile, in the Attic, Anahera and Nathan look down through the slit in the flooring. They see Max walking under the ceiling, opening and out into the cave.

With a flashlight in hand, Max walks a dozen steps through several inches of water until he can see to his right the end of the tunnel. A continuous light stream of water flows over parts of the black wall, but there are no paths beyond the side tunnel. He turns and makes his way back to the steel door. Konnie meets him, carrying a black leather bag, and they enter together.

“Anything?” Konnie asks.

“Nothing but black wet rock,” Max replies.

“What about the shaft in the warehouse?” Konnie inquires.

“It's filled with rocks,” Max responds.

Together, they walk to the workroom.

Konnie places the black bag holding the charges on the table. “Their Gator is still there. I found no sign of them on the hillside.”

“Then they have to be hiding somewhere in this mound of rock and mud,” Max deduces. Pick up the crystals and put them in the bag.

He takes out one of the charges and attaches it to the wall, as close to the ceiling as he can. Handing Konnie one of the charges, he removes the satellite phone from his utility belt and enters a code.

"Place this in the office, as high as you can at the far end," Max instructs.

Admiral Davis's voice comes over the phone. "Did you finish it?"

"Not yet. We know they're in here," Max replies.

"Did you check to see if there was a mine entrance?" Admiral Davis asks.

"Yes, but it was a shaft filled with rocks," Max responds.

"Then it must be entered from one of the other rooms or another entrance. That's where they're hiding," Admiral Davis says.

"Then I am going to level the rat hole and hope they come out before it comes down on them. Either way, it's their grave site," Max declares.

Admiral Davis's voice crackles through the phone. "Did you find the device or any files or anything else?"

Max glances around the debris-strewn room. "There

are only a few papers and some glowing crystal spikes, but that's about it."

"Damn!... Blow it!" Admiral Davis commands.

Max lowers the phone just as Konnie re-enters the room. "I set it for thirty minutes," she informs him.

Max hefts his M-16, eyeing the floor. "They may be in the caves below us," he suggests.

Max takes one of the charges, placing it on the wall near the ceiling, and hands Konnie another one. "Position this at the far end of the warehouse."

She takes the device and exits. Max walks across the broken glass door and enters the lobby.

"If you're down there... take a look around. It's about to be your grave," Max taunts, his voice echoing in the empty space.

Chapter 12:
A Long-Distance Call

Max sets his M-16 on the lobby counter and takes the last charge, moving to place it in the middle of the logo. Nathan slips down from the attic behind him. Max, not yet having set the charge, reacts to the sound just in time to meet Nathan's fist with his face. Nathan swings Max around and slams his head into the steel door. They wrestle across the room; Max desperately tries to retrieve his gun, pounding at Nathan's face, while Nathan delivers relentless punches to Max's chest and stomach. With a final, swift Jujitsu move, Nathan forces Max to the floor behind the counter in a chokehold.

Konnie enters the lobby with her Beretta ready, cautiously moving around the counter and under the attic box. Suddenly, she is ensnared by Anahera's legs, which have slid from the open attic box. Anahera's legs clamp around Konnie's neck, pushing her forward against the shattered glass. Konnie's face and body hit the glass hard, and for a moment, neither move.

Meanwhile, Max, still in the chokehold, manages to

pull a stiletto from his belt and jams it into Nathan's thigh. Nathan tightens the chokehold, grabs the blade, removes it from his thigh, and plunges it into Max's chest. Nathan bears down on the chokehold until he finally feels that Max is not moving. He pushes Max off him and looks over to see Anahera punching Konnie's bloody face twice, causing Konnie to fall backward. Nathan scrambles to his knees just as Konnie retrieves her Beretta. As she is about to aim at Anahera, Nathan grabs Max's M-16, brings it up, and blasts Konnie back into the workroom.

"You're bleeding!" Nathan exclaims, looking at the blood on Anahera's t-shirt and pants leg.

Anahera looks down at the blood. "I think it's hers. Is he..."

"Yes," Nathan confirms. "Are you sure you're okay?"

Anahera nods, not noticing the blood on Nathan's thigh. He stands and looks up at the hatch. "I'll get the sea bag and device down. Did you put the crystals in the picnic basket?" Anahera answers, "And two of the Plexiglas star balls, I had put them in the safe room. I'll get the basket.

Nathan moves the chair below the hatch and pushes up into the attic as Anahera heads into the office, careful not to get too close to Konnie's body.

Anahera retrieves the ledger and basket from the safe room and starts back toward the workroom. “Your leg!” she calls out when she enters and sees Nathan coming down from the attic.

“No time... he replies, “Take them to the beast!” Anahera hurries out of the office.

Nathan carries the container and the sea bag, stopping for a moment to look around. He notices a bright blue light reflecting on coal-black walls where Max shot up the area. Turning, hearing the buzzing of the satellite phone. At first, he intends to ignore it but then stops, Nathan picks up the satellite phone and presses the receiver button.

Admiral Davis’s voice demands, “What happened?”

“Who is this?” Nathan demands.

A pause follows before Admiral Davis's voice comes through, “Lieutenant. This is Admiral Davis, Naval Intelligence. Thank God you're okay. I was just informed that—”

“Save it, Admiral!” Nathan interrupts, his voice cold. “They didn't make it.”

Nathan hears loud noises over the phone and another

man's voice breaking through. “Put the phone down, Jack! I said put it down and don't disconnect. Admiral Jack Davis, I have a Federal warrant for your arrest.”

“Colin, I can exp—” Admiral Davis starts.

“Jack, take my advice and don't say another word. We have been listening to you for the last 24 hours. Take him away,” Colin Brown orders.

Nathan hears the commotion of Admiral Davis being taken into custody. Then Colin Brown's voice comes back on the line, “Lieutenant, Lieutenant...!”

Nathan remains silent.

“Don't hang up,” Colin Brown urges. “This is Colin Brown, NSA. We have just arrested Admiral Davis for crimes against the United States. Now, young man, I suggest you and Anahera get out of there fast. Take the stemmer ship that is coming to pick them up at the north pier.”

“Why should I trust you?!” Nathan demands.

“The way I look at it, besides the young lady and me, you don't have anyone in the world you can trust. Now, get the hell out of there! I will call you back,” Colin Brown replies firmly.

Nathan pockets the phone, grabs the device canister

and the sea bag, and exits the cave. He stops at Max's Gator just as Anahera pulls up. Thunder rumbles in the distance, and the sky darkens ominously.

“Change of plan. Follow me in the beast,” Nathan orders.

He gets into Max's Gator and floors it, with Anahera following closely. They speed through the jungle, navigating the rough terrain as fast as possible. They reach a clearing above the beach, and Nathan slides to a stop. Anahera pulls up beside him, the storm approaching quickly.

“Another storm is moving in,” Anahera observes.

They both turn to look back at the mountain. The ground begins to shake, and the top of the mountain collapses like a cheese soufflé. Smoke and dust drift over the area where the mountain once stood.

“Oh, I thought there would be more of... an explosion,” Anahera says, surprised.

“It was actually meant to be an implosion,” Nathan explains.

Anahera looks at Nathan, curiosity and determination in her eyes. “Well, Lt. Miller, what now?”

Nathan opens Jack's suitcase, finding a shirt, a pair

of pants, and shoes. He also spots a passport and what looks like ten thousand dollars wrapped in a one-hundred-dollar bundle. "That's up to you, Ana. I'm going to take the boat that was coming for those two nut heads and then..." He takes a handful of bills and hands them to her.

"This is your life and your home is on this island."

"And that's it?... Thank you little native girl, see you around?" Anahera lets the money fall back into the suitcase, moves closer to him, and looks deeply into his eyes. Suddenly, she makes a fist and slams it into his bloody thigh.

"Ouch!" Nathan cringes in pain.

"And who is going to bandage you up, every time you decide to take on someone with a machine gun...?!"

"I just—" Nathan starts, but Anahera cuts him off.

"...make you buttermilk pancakes, find you a bag of coins, gold bars, hundred-dollar bills and, by the way, I understand it is finders keepers, but I am willing to split it with you and also..."

He puts his arms around her and interrupts her words with a long kiss… then… "So, you're coming with me," he says, still holding her.

She slowly moves away from him and turns Konnie's

suitcase around to open it. “Well, that depends on what she has to wear in here.” Anahera looks at the contents and exclaims, “Ohhhh! I think wherever we end up, I'm going to need to go shopping for something other than black.”

Anahera continues to look at some of the black lace items in the suitcase. Nathan smiles, “I like black.”

She removes a black scarf, and a small green package from the case and closes the suitcase. She looks at Nathan's bloody pants. “Now, take off those pants, and let me look at what you've done to yourself this time.”

Nathan slips off his pants while Anahera removes a leaf from a vine and pours the water from it on his wound. “Incredible! It's already healing.” She opens the small package, removes the pad, places it on the wound, then wraps the scarf around his thigh and pulls it tight.

“What about the clinic?” Nathan asks.

Anahera starts getting dressed while Nathan takes water from another leaf, wets the t-shirt he took off, and places it on his face. Nathan wipes his hands and face. Anahera turns her back to him, slips off the Navy t-shirt, and steps under a large leaf. The water falls on her head and body. “As far as they are concerned, I left the island yesterday.”

Nathan gets dressed, takes Max's passport and wallet with a few hundred dollars in it, places them in his pocket, and puts the rest of the cash into Konnie's suitcase along with a pair of Max’s jeans and a colorful short-sleeved island shirt. He closes Max’s suitcase and tosses it into the heavy brush.

“If you don’t show up in a few days...”

“They don't expect me back here for another week. I will just let them know that I have decided to do some traveling.” She has slipped on a pair of very tight, black satin pants and a black blouse.

The satellite phone buzzes. At first, Nathan lets it buzz, but it just keeps repeating itself. He picks it up.

“You made it out of there. Is Anahera okay?” Colin Brown's voice comes through.

“Look, I'm not coming back. I don't—” Nathan starts.

“I understand that, and I don't blame you. I am sorry that your life has been turned upside down,” Colin interrupts.

Nathan cuts off Colin Brown mid-sentence. “I no longer have a life. That was taken away from me by some crazy, insane experiment!”

“You're right,” Colin responds. “And as far as your

government is concerned, you were shot down. You had already made ace, and in my book, you're a hero."

"And a lab rat. I will not come home," Nathan insists.

"Okay, look at it this way. Your government owes you over eighty years back pay," Colin offers.

"So...?" Nathan asks, his voice tinged with skepticism.

"I want you to understand that I'm sorry and, if you'll trust me, I will help you," Colin says earnestly.

"Now we are back to trust," Nathan says, a note of disbelief in his tone.

"Yes. I plan on earning that trust," Colin assures him.

"I'm listening," Nathan concedes.

"The steamer Max booked is safe. The company checked out. They were going to Mindanao," Colin explains.

"Why Mindanao?" Nathan queries, his curiosity piqued.

"Seems you returned fire and hit the windshield, splintering the woman's face," Colin reveals.

Anahera nods, "That's true. Her face and arm were bandaged."

"There is a well-known plastic surgeon in South Cotabato Mindanao there who works off-book," Colin continues.

"So, what am I supposed to do?" Nathan asks, feeling a sense of direction amidst the chaos.

"It's your life; do whatever you decide. I'm sure you found the money that was left behind, As well as a few other items, Colin says."

"And...?" Nathan prompts.

"I will give you the name of a banker near South Cotabato who will exchange the old money for new currency," Colin offers.

"More trust?" Nathan asks, raising an eyebrow.

"It's your choice," Colin states simply.

"I'll think about it," Nathan says, weighing his options.

"Fair enough. I will call you in the morning. Oh yes, this phone is now encrypted to only my number and voice," Colin concludes.

Nathan lowers the phone. Anahera looks at him, concern in her eyes. "Who was that?"

“A government spy agency called the National Security Agency; they just arrested the Admiral,” Nathan explains.

“I have heard of it on the national news, NSA,” Anahera says, her brows furrowing.

This guy was responsible for the Admiral's arrest and knew all about me and apparently you,” Nathan continues.

“What does he want?” Anahera asks, her tone wary.

“To help us,” Nathan replies.

“Can he be trusted?” Anahera questions.

“We may know by tomorrow,” Nathan says, a note of finality in his voice.

A beat passes. Anahera speaks softly, “She had a bandage on her face and her right arm. I think I can come up with something until we are aboard.”

Ana opens Konnie's suitcase again, removes a lacy blouse, rips the bottom lace, and places it on her head and the left side of her face. She then takes out a stylish black hat and puts it on her head. “Something like this?”

Nathan replies, “Should work. Dump what you don’t need and put the contents of the picnic basket in it.”

Anahera looks at Nathan through the lace, a playful glint in her eye. “I kind of like this style.”

Nathan climbs on the beast. “I'm going to get rid of this thing.”

“Just take it down to the beach. We can take this one to the pickup point at the North Pier,” Anahera suggests.

A blast of thunder and lightning strikes, touching the sea in the distance, underscoring the urgency of their situation.

***

Aboard the four-cabin steamer ship, Anahera and Nathan enter the small cabin and place the suitcase, sea bag, and device canister in the closet. Anahera takes off her hat and scarf and enters the very tight bathroom to wash her face. The storm has arrived, and the seas are angry, causing the steamer to sway harshly with the movement of the wind and current.

***

At the jungle’s hillside edge above the beach, a separate gaiter pulls up and stops. Two men in raincoats look out at the steamer that has headed out into the churning sea. One man quickly yells out, “I don’t think that was Max.” The other man starts the gaiter. “If not, it could make our job easier. Right now, we have to find a way to get to Mindanao as fast as possible.

Chapter 13:
The Friendly Banker

As Anahera and Nathan sit down to a light breakfast at the small seaside hotel restaurant near South Cotabato, Mindanao, they are still wearing the clothes of Max and Konnie, their aliases. The white bulging sea bag stands next to Nathan's chair, a constant reminder of their precarious situation.

"I'm still not sure this is a good idea," Nathan says, his brow furrowed with concern.

Anahera reaches out and takes his hand in hers, her touch warm and reassuring. "He got us through customs with these passports and called, as promised, with the name of the banker," she reminds him.

Nathan's eyes narrow. "I know, but..."

Anahera squeezes his hand. "I have a good feeling about this man. He sounds sincere, but, as he said, it is your life."

Nathan's gaze meets hers, and for a moment, they just look at each other. "Not anymore," he says finally. "It's our life."

Anahera's face softens, and she gives his hand another gentle squeeze. "We can leave right now, if you want."

Nathan hesitates, then nods. "You have to promise me that if I am not back in an hour and a half, you'll check out of the hotel, take the money in the suitcase, and get off this island. Understand?"

He takes her hand in his and squeezes it, his eyes locked on hers. Anahera nods, her face set with determination. "I promise. Just be careful."

Nathan stands, shouldering the sea bag, and gives Anahera a quick kiss. "I will," he says, then turns and walks away, leaving Anahera to watch him go with a mix of worry and concern etched on her face.

Twenty minutes later, a taxi pulls up in front of a large building. Nathan exits and walks into the Mindanao World Bank. The interior of the business bank lobby is a hive of activity. A series of desks stretch out before him, each one occupied by a banker chatting with a customer. At the far end of the room, a small teller window is situated next to two large carved wood doors. Near the front door, a small guard's station sits behind a one-way glass window.

Nathan's eyes scan the room, his senses on high alert as he makes his way through the Lobby. He has no idea what to expect, but he knows he needs to be careful.

Nathan approaches the station, stopping in front of the glass to glance at his own reflection. “I am here to see Mr. Wentworth,” he says, his voice firm and confident.

Before he can finish his sentence, an older, attractive native woman, Miss Raira Agbayani, walks up to him. “Mr. Wentworth is expecting you. Please come this way,” she says, her smile warm and welcoming.

Nathan follows her to the double doors, which open just before they reach them. She leads him into a large, well-appointed office, the décor a blend of modern and traditional.

“Send in the young man,” a deep voice says from the box on her desk.

The doors open, and Nathan, with the sea bag still slung over his shoulder, enters. Thomas Wentworth, a good-looking, tall Filipino gentleman in a crisp white suit and open collar, greets him from behind a giant desk. To his right, a glassed-in room holds a conference table, while behind him, two dozen or so monitors display every room in the bank, as well as the four outside views and roof.

Thomas extends his hand, and Nathan responds with a firm handshake. “Please sit down,” Thomas says, his smile warm. “I take it that what you have brought is for me to evaluate?”

Nathan nods, taking a seat in one of the plush chairs in front of the desk. “Yes, sir.”

Thomas's eyes crinkle at the corners as he smiles. “It's Thomas, please. Before we get to business, would you like a drink or some Fiji water?”

Nathan shakes his head. “No, thank you.”

Thomas nods and hands Nathan a carved leather box. “Our mutual friend asked that I have these prepared for you. He said he would explain later.”

Nathan opens the box, his eyes widening as he takes in the contents. Four passports, a booklet, and some other documents lie inside. “These were prepared overnight?” he asks, his voice laced with surprise.

Thomas gets up from his chair and starts toward the conference room, his movements fluid and confident. “Let's take a look at what you've brought, shall we?” he says, his voice low and smooth.

As Thomas Wentworth leads Nathan into the

conference room, Nathan's hesitation is still palpable. But he follows Thomas, loosening the ties on the sea bag and dumping its contents onto the conference table.

Thomas's eyes light up as he surveys the 1941 one-hundred-dollar bills, wrapped in bundles. Some are stamped with the word "Hawaii" front and back. Thomas picks up the ones over-printed with "Hawaii", his fingers moving with a practiced ease.

"These are very rare," Thomas says, his voice low and smooth. "Is this all of it?

Nathan nods, his eyes fixed on the bills. "Yes."

Thomas looks up at him, a calculating glint in his eye. "Well, son, how about we settle on say five million?"

Nathan's gaze flicks up to Thomas's face, then back down to the bills. "Five million...cash?"

Thomas nods, his smile growing. "Cash, if you like, or I can deposit some and—"

"Cash would be fine," Nathan interrupts, his voice firm.

Thomas nods again. "United States currency in one-hundred-dollar bills."

"Now?" Nathan asks, his eyes locked on Thomas's.

Thomas nods once more. “Yes, in a few moments. In the meantime, a drink?”

Nathan hesitates for a moment, then nods. “Water sounds good.”

Thomas turns and opens a wood box on a long table, removing two Fiji bottles and setting them on crystal coasters on the table, along with two cut crystal glasses. “Please sit down,” he says, his voice warm and inviting.

Thomas pours the water into the glasses, his movements fluid and precise. Nathan takes a seat, his eyes fixed on the glasses.

As Thomas hands him a glass, Nathan speaks up. “If it's possible, I would like three million placed in this.”

He hands Thomas the sea bag, and Thomas pushes a button on the box in the middle of the table. He speaks in Chavacano, and in seconds, the office door opens and his assistant, Miss Agbayani, enters. She picks up the sea bag without saying a word and walks out.

Thomas turns back to Nathan, his smile still in place. “Do you plan to stay in the city long?”

Nathan takes a drink of water, his eyes fixed on Thomas. “Not sure yet.”

Thomas nods, his expression understanding. “That is a lot of cash to be walking around with in this city. I have several executive suites at the Royal Aster Hotel. I would like you to be my guest. There are safes in the room and excellent security.”

Thomas reaches into his inside coat pocket, takes out a gold case, opens it, and removes a card. “Give them this, and I will have Miss Agbayani call.”

Nathan takes the card, his eyes flicking over it. “We're not sure yet what the plan is.”

Thomas takes a drink of water, his eyes never leaving Nathan's face. “For your safety, I highly recommend it. Is the young lady close by?”

Nathan hesitates for a beat and answers, “Yes.”

The doors open, and a gentleman enters carrying a black, aluminium case and the sea bag. The man opens the case for Nathan to see the contents, then shuts it and hands Nathan a small gold case and exits. Nathan looks at the gold case, his eyes narrowing.

Thomas smiles. “It's the code for the case. It's bulletproof and vertically indestructible.”

Nathan's gaze flicks up to Thomas's face. “Thank

you."

Thomas nods, his smile growing. "Don't forget your passports. I will have Miss Agbayani call ahead, and there should be a limo here momentarily. Just in case."

Thomas escorts Nathan to the doors.

As Nathan exits the conference room, Thomas calls out to him. "Be safe, my friend. You have my number if you need anything else."

Nathan nods, then disappears from view. Thomas stands and watches one of the monitors, tracking Nathan's progress as he walks through the bank. Then, he picks up the phone.

"Yes, he is just leaving," Thomas says into the phone. "He came in alone."

A beat passes, and Thomas continues, "No problems. I believe he was satisfied with the transaction. He seems like a good kid, but we both know his troubles are far from being over. I am now sure he has no idea what he soon will be faced with."

Another pause, and Thomas says, "Around eight million... I gave him five. He didn't bring in any gold."

Thomas's voice is smooth and calculating.

"Expenses, my friend. Those rushed passports were not cheap."

Thomas concludes, "I understand. I'll have my people keep an eye on him while he is here on the island."

Nathan sits on a super king-size bed in a plush hotel suite surrounded by the trappings of luxury. Just outside the wide windows, a private pool glimmers in the afternoon sun. The room appears to be on the first floor.

He is wearing a pair of Max's jeans and a colourful short-sleeved shirt with lower pockets on each side. He has the leather box open, revealing the new passports and a booklet containing his and Anahera's dossiers with their new identities, as well as his multi-engine pilot's license.

He reads the name on the dossier, "Nathan Falcon. Well, better than Smith or Jones, Colin."

He looks at Anahera's passport and reads the name, "Angela Falcon. You definitely have been reading too many comic books, Colin."

He places the documents back into the leather box, opens Konnie's suitcase, looks at the Plexiglas star balls, and removes one of the dusty charts found in the secret room. He studies it for a moment, then says, "A self-contained unlimited power source." At the top of the chart are the words "The Nikola Tesla Factor." He continues, "Unlimited single power source without an exterior generator. I keep saying it's impossible, but we are carrying around the proof in a suitcase."

Nathan hears something and points his automatic at the door.

The door swings open, and Anahera enters, pulling a wheeled suitcase behind her. Nathan's automatic is pointed at the door, but he lowers it as soon as he sees Anahera. She closes the door and sets the luggage by the bed, then performs a 360-degree turn, showing off her new colourful slacks and blouse.

"What do you think?" she asks, her eyes sparkling.

Nathan closes the leather box, puts down the chart, and smiles, taking in her new outfit. "You look amazing."

Anahera grins, her excitement palpable. "I have had the most exciting and enjoyable afternoon of my life." The hotel limo took me to the local stores and the driver escorted me everywhere.

"The luggage?" Nathan asks, curiosity getting the better of him.

Anahera launches into a breathless explanation. “I bought this great luggage set first, and then went into a marvellous shopping complex and bought each of us four full sets of clothing. Every item I found, I immediately packed in the luggage. We can now dump all of these strange outfits.”

Taking a breath, she smiles at Nathan with a look of satisfaction.

Nathan holds up the leather box. “Our new passports, a gift from Colin.”

He reaches for Anahera and pulls her over next to him. He looks down at her, his eyes filled with admiration.

“You are, by far, the most incredible woman I have ever known,” he says, his voice low and husky.

Anahera teases him, her eyes sparkling. “Have you known a lot of young women?”

He moves closer, and they kiss, the tension between them palpable.

The satellite phone buzzes, breaking the spell. Nathan sits up and picks up the phone, his eyes never leaving Anahera's face.

Chapter 14:
Nikola Tesla's Dream

Nathan sits up, his heart racing as he picks up the phone.

"Colin Brown," the voice on the other end says. "Nathan, Thomas is sending someone to get you out of there and to the airport."

Nathan's eyes widen in surprise. "What?"

"There is another shooter," Colin explains. "Go! I will explain on the way to the airport."

Just then, there is a knock at the door. Nathan swings off the bed, his automatic pointed at the door. Anahera moves off the bed and stands behind him, her eyes fixed on the door.

A voice from the hallway outside the door calls out, "Wentworth sent me."

Nathan cautiously opens the door, his weapon still trained on whoever enters. A man stands in the doorway, his eyes scanning the room.

"Mr. Wentworth sent me," Narito says. "You must leave immediately. I am here to take you to the airport."

Nathan lowers his gun, his grip relaxing slightly. “Ana, the suitcase with the crystals,” he says, grabbing up the canister.

Anahera quickly complies, grabbing the box on the bed and Konnie's suitcase as Nathan throws the sea bag and canister on top of the two new luggage bags. Narito has already started towards the door, a Glock in his free hand. Nathan picks up the black money case, and the three of them move quickly out of the room.

As they reach a door at the end of the hall, Narito opens it to another hallway. “The door on the left leads to the back loading area, where there is a black Mercedes high-roof passenger van waiting,” he explains.

As they reach the door, Narito turns and fires past Nathan and Anahera. A man takes a bullet from Narito's Glock.

“They're here,” Narito says, his voice urgent. “Go! Hurry!”

Nathan and Anahera move quickly through the door, Narito waiting a moment before turning and pulling the luggage behind them. They find themselves in a dimly lit loading area, a black Mercedes high-roof passenger van waiting for them. A man inside helps Nathan put the canister and black case into the rear area as Anahera gets into the back seat. Narito runs up, tossing the luggage into the side area, before getting in beside

Anahera. Nathan follows, the other man now in the driver's seat. The van pulls away, speeding down the road behind the hotel and turning onto a dirt road.

The van skids around the corner, dust flying everywhere as they speed away. Nathan looks back, seeing two men emerge from the back door, guns in hand, watching as the van disappears into the distance. A black SUV pulls up to the loading area, and the men climb in, the SUV now in pursuit.

Inside the van, the windows are dark, blocking out the view, but Nathan can feel the van's speed and agility as it navigates the dirt road. Anahera's hand reaches out, grasping his, her eyes fixed on his.

"What's going on?" she asks, her voice barely above a whisper.

Nathan shakes his head, his eyes fixed on the back of the driver's seat. "I don't know, but we need to get out of here fast."

The van takes a sharp turn, throwing them against the side of the vehicle. Narito's Glock is still clutched in his hand, his eyes scanning the road behind them. He hands Nathan a box of .45 caliber bullets. "I figured you might need these." Nathan takes the box and empties the shells into the side pockets of the island shirt.

“We're not out of this yet,” he mutters, his voice low and deadly serious.

Anahera's arms wrap tightly around Nathan as she asks, “What's happening?”

The satellite phone buzzes, and Nathan picks it up. “Colin Brown,” the voice on the other end says.

“Sorry about that,” Colin apologizes. “Are the two of you okay?”

Nathan's voice is laced with frustration. “What the hell?”

“We were able to break the Admiral's inscribed phones,” Colin explains. “He had sent another shooter to kill the first two and you and grab the device. They obviously got there as the boat was pulling out into the storm.”

Nathan's mind is racing. “Now what?”

“Thomas' men are taking you to a small airport where he has a plan ready to get you off the island,” Colin says. “Thomas' men will take care of these hit men. You just get off this island.”

Nathan presses for more information. “Then?”

“It's your decision,” Colin replies.

Nathan remarks with a touch of sarcasm, “You keep saying that, but so far, none of this has been my decision.”

Colin continues, “Unfortunately, it's partly my fault. When you inquired about your past and showed up in our office, I remembered Admiral Davis investigated these strange types of phenomena a number of years ago, so I called him in to review the file. I had no idea the Admiral was involved in this ghost of the past for years, so, as I said, I walked us both right into this.”

Nathan's curiosity gets the better of him. “Why are you doing this? You could have just handed me over to the CIA or whoever, and that would have been the end of this.”

Colin's response is simple. “With what I know now, that is not the way it would have gone down. The moment I brought the Admiral in on this, you were a dead survivor of a war over eighty years ago—a man who now knows too much about a government cover-up. As I said, you're a special kind of hero, and right now this country and this situation need you. I am sure by now you know there is a lot more to this than just a device you were chosen to deliver.”

Nathan's eyes narrow. “The power crystals, the Nikola Tesla Factor?”

After a moment, Colin confirms, “The very thought that there may be an infinite, simple source somewhere out there to generate unlimited power for the world is a frightening thought. No individual or country, under any circumstance, can discover or control the truth that you possess—a truth that has already taken too many lives and, until now, remained a ghost of the past in two unfortunate, ill-fated, murderous air crashes that lie crushed at the bottom of the South Seas. Tesla predicted that 'no free energy device will ever be allowed to reach the market.' A reality that must, for now, remain in the lava tunnels below a small resort island in the South Pacific. Fortunately, the Admiral was unaware of the full meaning of the Nikola Tesla Factor. He related it to the device, not understanding the power within Professor Augustus Franco's true discovery."

Nathan asks, "Then you do know the truth?"

Colin's voice crackles through the phone, "Once you're in the air and safe, I'll try to track you, and we can talk then."

The call ends, and Nathan lowers the phone, his eyes locking onto Anahera's, who tightly grips his hand.

The van speeds through the curves, skidding onto a paved road beside a tall wire fence. The driver's skilled hands navigate the twists and turns, finally approaching the airport gate. A guard opens the gate, and the van passes through onto the airfield.

The CESSNA 310P, with turbocharged Continental TSIO-520 engines, is already ignited, the pilot pulling down the passenger seat. The large limo van provides cover as the driver hurriedly loads the luggage and sea bag behind the third seat. Nathan helps Anahera onto the wing and into the rear compartment seat. He moves back to the van to retrieve the black case.

Meanwhile, back at the airport gate, the SUV that had been following them skids up to the gate. The driver yells at the guard, “Open the gate!” The guard responds, “This is a private air strip; back away!” The driver pulls out a gun, and the guard reaches for his own. The driver fires, knocking the guard back against the gate. The guard returns fire, hitting the driver in the neck. The guard falls to the ground, and the driver slumps forward.

Narito turns around, an M-16 in his hand, and is joined behind the van by Nathan, who has his automatic. The van driver has finished loading the plane.

The air is thick with tension as they prepare to take off, leaving the danger behind.

As the SUV's passenger, Poga, exits the vehicle, he's followed by Crago from the back seat, M-16 in hand. Crago's orders are clear: “Get him out of there and get the damn gate open!”

Poga makes his way around the back of the SUV, moves to the driver's door, opens it, and pulls the driver out, dropping him on the ground. Meanwhile, Crago raises his weapon from behind the right front door and fires several rounds.

Nathan and Narito return fire from behind the van, the sound of gunfire filling the air. Narito glances towards the gate house, his eyes fixed on the target.

Poga reaches the gate house, but his progress is short-lived. He takes several bullets and falls to the gate house doorway, dead.

Crago's scope is trained on the Mercedes van, slowly moving past it and then up to the Cessna's cabin. The pilot turns to yell something, but Crago's bullets tear through his flying jacket, entering his shoulder just below the collarbone.

Anahera's voice rings out, yelling to Nathan, “The pilot has been shot!”

Narito's response is firm, “Stay down!”

The exchange of fire continues, with Crago dropping down to replace his empty clip. He slowly comes up firing, shattering the front driver's side window, but the bullets have no effect.

Inside the cabin, Anahera lowers the pilot's seat so he's partly in the back seat.

Narito yells to Anahera, “Can you fly that thing?”

Nathan's response is immediate, “Yes!”

“I'll give you cover, then get the hell out of here!” Narito replies.

“What about you?”

“Just go now. We will meet again.”

With that, Narito stands and continues to blast Crago's car door, providing cover for Nathan's escape.

Nathan moves up the wing, slides into the right seat, and immediately powers up the engines.

Awang Tower, 247 Echo Fox Departing 28 North West.

The radio voice crackles, “247 Echo Fox, you are cleared for departure North West.”

The Cessna begins to taxi and pick up speed, leaving the chaos behind.

Crago drops down behind the door, momentarily escaping Narito's hail of gunfire. But then he pops up, aiming his weapon at the Cessna. Narito continues to fire, trying to distract Crago, but the latter is undeterred.

As the wheels of the Cessna are about to lift from the airstrip, Crago fires and then steadies his weapon to fire again. But in that moment, a bullet finds its mark, and the hair on the back of Crago's head blows back. A piece of his skull flies out, turning the air red.

In the cockpit of the Cessna, Anahera is tending to the wounded pilot. She has torn the pilot's shirt away from the wound and has taken a handful of tissues from the box attached to the back of the right seat, placing it on the wound. She has tied the tissues in place with the pilot's belt.

"How's he doing?" Nathan asks, his voice laced with concern.

"He has lost a lot of blood," Anahera replies, her voice steady. "He needs to get to a hospital."

The satellite phone buzzes, and Nathan picks it up. "Colin Brown," the voice on the other end says. "What's your situation?"

Nathan's eyes are fixed on the instrument panel. "We are at five thousand feet, heading northwest and out of there. The pilot took a bullet, and we've got to get him to a hospital."

Colin responds, "Give me a moment." The line goes silent for a beat, and then Colin's voice returns. "Is there a chart?"

Nathan pulls one down from the visor. "Yes."

Colin's voice continues, “Can you make it west-northwest to Lebak Airport?”

Nathan searches the chart. “Got it.”

Colin continues, “They will be expecting you. An ambulance will meet you at the far end of the dirt runway… Awang UPI gate. As soon as he is out and safe, take off and get the hell out of there. You have everything you need to start a new life. Leave the ghost of the crystals buried.”

Nathan, with a curious tone, asks, “You just called an airport in Mindanao and cleared us to drop off a wounded Filipino pilot, then just take off?”

Colin answers, “Yes. Remember, we are the NSA. Awang Tower has you on radar; there is no Tower at Lebak. You don’t need a clearance; you should land in fifteen minutes. Nathan, are you there? Nathan?”

Nathan puts the chart down and makes a slight adjustment. He answers, “Yes.”

Nathan looks down at the pilot, who is in pain and tries to smile as he looks up at Nathan. “Thank you.”

The plane begins its descent toward the airstrip in the distance.

Chapter 15:

It's not all about fiction

The morning sun rises over the ocean's horizon, painting a large beach cottage with an orange glow. The two-story building sits on a hill overlooking the South China Sea. A blue-trimmed Cessna 421 C Golden Eagle sits nearby on a narrow runway on the cliff's edge.

Inside the beach house, Nathan sits in a leather chair at a desk in front of a large monitor. He wears a headset with an attached microphone. As he speaks, the words appear on the screen at a rhythmic pace. Next to the monitor is a model of a 1943 Grumman F6F Hellcat. The desk is surrounded by floor-to-ceiling bookcases. A rocking chair is the only other piece of furniture in the small room.

The words on the monitor appear as follows:

As the Lieutenant opens his eyes and looks around the dimly lit room, he realizes he is lying in a hospital bed. His wrists are bound, and there is a needle attached to a plastic tube traveling up to an IV bag above his head. A heart monitor pulsating his vitals continues to echo. He can make out the voices of two men talking somewhere nearby.

The words flowing onto the screen as his voice continues to tell the story.

A young man with a Filipino accent addresses another man. “Right now, General, the lieutenant believes he is the only survivor of a DC6 he crashed into the China Sea, causing the death of eleven passengers.”

The General, who has a more distinctive manner in his words, replies, “Doctor, the mind-altering drugs you have given him, are you sure they will permanently erase all his memory of the past.”

Their voices seem to be growing louder, as the young man pulls on the wrist straps that bind them to the bed rail, then closes his eyes and continues to listen as the men enter the room.

The Doctor approaches the IV pole next to the bed and carefully adjusts the constant drippings within the small tube that is regulating the flow from the IV bag of liquid into the tube that is attached to the patient's arm. “Yes, every drop that enters his veins continues to block his memory up until what he believes was the crash. We will then slowly rebuild his memory as commercial pilot Nick Porter. That will...”

The General cuts him off sharply, “Why not just put a bullet in his brain and end this?!”

The Doctor is a bit uneasy, but responds while he remains at the bedside with his back to the General. “No, as long as he believes he was the cause of the death of the men in the crash, we can hold him and continue to study how he developed all these extraordinary powers.”

The General obviously not convinced with the doctor's answers, “What about the girl?”

The Doctor turns and looks to the General. “We will hold her here until we're sure we have made the conversion, then dispose of her.”

The General responds indignantly, “Again, Doctor, understand once you do this, he must not have any memory of the blue crystals, the caverns, or the island.”

The Lieutenant has heard enough and rips the hand zip tie off his wrist...

As the words continue to appear on the screen, a voice is heard, “Nathan, come on. Your pancakes are going to get cold.”

Nathan continues to dictate.

The Lieutenant pulls the needle out of his arm and bounds off the bed, wearing only a hospital gown. He comes face to face with a three-star General and a young man in a white coat.

“Nathan!” Anahera's voice calls out again.

Nathan pauses the headset. He stands and takes it off, setting it on the desk.

“Coming,” Nathan replies loudly.

He exits the room, leaving the words on the screen.

In the kitchen, Ana is waiting for him, wearing shorts and a halter top covered by a colorful apron. She places a plate down with three pancakes piled high, and Nathan sits down.

“The publisher sent the books for you to sign for Sunday,” Ana says.

“Great,” Nathan replies.

On the kitchen counter, a partially opened package holding several books sits with other mail. A light blue flyer is clearly visible on the wrapping. Gold letters appear over a photograph of a crystal with a ten-pointed star on a soft blue background. Printed on the flyer are the words: “From another bestseller, 'NAT FALCON'S CHRYSTAL BLUE, now comes the latest South Sea thriller, CHRYSTAL CAVERNS, on shelves today.”

Ana sits down with a cup of coffee as Nathan takes the first bite of his pancakes.

"How's it coming?" Ana asks.

"Okay. Roughing out the hospital wake-up segment," Nathan replies. "Did a lot of research this morning before light."

***

At that very moment, on a rocky ledge at the north edge of Half Moon Bay, over 800 miles south of the rustic beach cottage, a fifteen-foot wave crashes violently against the shoreline. For a brief moment, the waves pull back, exposing a cave cut into the rocks, followed by another wave and then another. Suddenly, the sea calms, revealing a glowing Plexiglas sphere about the size of a basketball. Within the glowing sphere, a ten-pointed crystal star rests on top of a second crystal star, forming a double pentagram. It bobs up and down in the current as it is pulled out to the open sea, away from the island.

***

Inside the cottage, a familiar buzzing sound can be heard as Nathan sets down his fork. He and Ana exchange glances.

In the bedroom, a satellite phone sits on a table next to a blue vase with a bouquet of blue asters and a framed photograph of Nathan and Ana. The buzzing light continues to blink.

THE END

The ghosts from the power crystal lava caves have returned.

Navy Aviation 1941 to 1945

Excerpts from President Roosevelt's Speech Delivered on 8 December 1941

"Mr. Vice President, Mr. Speaker, members of the Senate and the House of Representatives: Yesterday, December 7th, 1941 — a date which will live in infamy — the United States of America was suddenly and deliberately attacked by Naval and Air Forces of the Empire of Japan. The United States was at peace with that nation and, at the solicitation of Japan, was still in conversation with its Government and its Emperor looking toward the maintenance of peace in the Pacific. Indeed, one hour after Japanese air squadrons had commenced bombing in the American island of Oahu, the Japanese Ambassador to the United States and his colleague delivered to our Secretary of State a formal reply to a recent American message and while this reply stated that it seemed useless to continue the existing diplomatic negotiations, it contained no threat or hint of war or of armed attack. It will be recorded that the distance of Hawaii from Japan makes it obvious that the attack was deliberately planned many days or even weeks ago. During the intervening time, the Japanese government has deliberately sought to deceive the United States with false statements and expressions of hope for continued peace.

The attack yesterday on the Hawaiian Islands has caused severe damage to American naval and military forces. I regret to tell you that very many American lives have been lost. In addition, American ships have been reported torpedoed on the high seas between San Francisco and Honolulu.

Yesterday the Japanese Government also launched an attack against Malaya. Last night Japanese forces attacked Hong Kong. Last night Japanese forces attacked Guam. Last night Japanese forces attacked the Philippine Islands. Last night the Japanese attacked Wake Island. And this morning the Japanese attacked Midway Island.

Japan has, therefore, undertaken a surprise offensive extending throughout the Pacific area. The facts of yesterday and today speak for themselves. The people of the United States have already formed their opinions and well understand the implications to the very life and safety of our nation. As Commander-in-Chief of the Army and Navy, I have directed that all measures be taken for our defense. With confidence in our armed forces—with the unbounding determination of our people—we will gain the inevitable triumph—so help us God. I ask that the Congress declare that since the unprovoked and dastardly attack by Japan on Sunday, December 7th, 1941, a state of war has existed between the United States and the Japanese Empire."

The United States entered World War II on December 11, 1941, declaring war on Germany, which had already been at war with the United States' allies since September 1939, when Germany invaded Poland, France, and Great Britain.

In December 1941, Navy aviation had only 3,400 aircraft, 4,600 pilots, and 1,009 commissioned officers. The average age of a pilot was between twenty and twenty-two years old. During World War II, the United States Navy trained pilots in Pensacola, Florida, or Corpus Christi, Texas. The aviation cadets trained for six months before being commissioned and assigned to a squadron. At the beginning of the war, flight training lasted nine months, divided into three months each of primary, basic, and advanced training. Each pilot had sixty-five flying hours of primary training and seventy-five hours of both basic and advanced training. The Flight Officer Act of 1942 created the warrant officer rank of Flight Officer.

During the first part of 1941, new pilots learned how to fly the various available airplanes. Navy fighter pilots would face the Japanese Mitsubishi Zero, a treacherous, nimble fighter plane. It wasn't until the Navy pilots received aircraft like the Corsair and a navalized version of the P-51 Mustang, called the ETF-51D for land-based operations, that the tide began to turn in favor of the Americans against the Zero. There is no question that from 1941 to 1945, the highly dedicated Navy pilots wreaked havoc in the Pacific theater. The Navy airmen of WWII faced their call to duty and are remembered for their relentless courage and sacrifice.

The death of Adolf Hitler by suicide in a Berlin bunker was revealed on April 30, 1945. On June 5, 1945, the final signatures on the instrument of Nazi Germany's unconditional surrender brought the war with the Rome-Berlin Axis to an end.

Three months later, on August 6, 1945, The Enola Gay, a Boeing B-29 Super fortress bomber named after Enola Gay Tibbets, the mother of the pilot, Colonel Paul Tibbets, took off from North Field in the Marianas. At 8:15 a.m. local time, The Enola Gay dropped the world's first atomic bomb over the Japanese city of Hiroshima. Sixteen hours later, American President Harry S. Truman called again for Japan's surrender, warning them to "expect a rain of ruin from the air, the like of which has never been seen on this earth." Late on August 8, 1945, in accordance with the Yalta agreements but in violation of the Soviet–Japanese Neutrality Pact, the Soviet Union declared war on Japan. Soon after midnight on August 9, 1945, the Soviet Union invaded the Japanese puppet state of Manchukuo. Hours later, the United States dropped a second atomic bomb on the Japanese city of Nagasaki. Emperor Hirohito ordered the Supreme Council for the Direction of the War to accept the terms set down by the Allies in the Potsdam Declaration. After several days of behind-the-scenes negotiations and a failed coup d'état, Emperor Hirohito gave a recorded radio address across the Empire on August 15th, announcing the surrender of Japan to the Allies.

World War II would be inscribed as the deadliest military conflict in world history, with seventy to eighty-five million people killed—about 3.76% of the world's population at the time.

www.ingramcontent.com/pod-product-compliance
Lightning Source LLC
LaVergne TN
LVHW022057200926
848481LV00003B/955
9798330491193